DEFENCE INDUSTRIES: A GLOBAL PERSPECTIVE

DEFENCE INDUSTRIES

A GLOBAL PERSPECTIVE

DANIEL TODD

ROUTLEDGE
London and New York

First published in 1988 by
Routledge
a division of Routledge, Chapman and Hall
11 New Fetter Lane, London EC4P 4EE

Published in the USA by
Routledge
a division of Routledge, Chapman and Hall, Inc.
29 West 35th Street, New York NY 10001

Printed and bound in Great Britain by Mackays of Chatham PLC, Chatham, Kent

British Library Cataloguing in Publication Data

Todd, Daniel
    Defence industries : a global perspective.
    1. Armaments industries, to 1986
    I. Title
    338.4'76234'09

ISBN 0-415-00411-X

CONTENTS

List of Tables
List of Figures
Acknowledgements

TABLES

FIGURES

ACKNOWLEDGEMENTS

I am indebted to the host of observers, commentators, pundits and journalists who labour at the daily task of disclosing facts and propagating opinions about the defence industry. They serve, in short, to cast light on an area shrouded in government obfuscation and secrecy as well as the usual reticences associated with any information deemed vital to corporate and national rivalries. Without access to their work in specialised journals, trade magazines and other releases, a study of this kind would be nigh on impossible. To my academic colleagues, who alerted me to the strengths and weaknesses of 'academic' inquiries into the defence arena, I am also grateful. In particular, I wish to mention Jamie Simpson, now at the London School of Economics, for his insightful approach to the whole question of the relations between defence and economic development. At a more practical level, I cannot overlook the unstinting 'back-up' provided by Marjorie Halmarson, Evelyn Gaetz and Jill Michalski in the draughting and typing that goes into manuscript preparation. Last, but not least, I owe much to my wife and son for bearing with me throughout the duration of the venture.

Chapter One

ARMS SPENDING AND DEFENCE INDUSTRIES

Before embarking at the beginning, that is, with a discussion of general patterns of international defence procurement, a few qualifications are in order concerning the value of military expenditure data. First and foremost is the lack of standardisation of national data and the problem of comparability thereby posed: a problem, incidentally, which has heretofore remained unresolved despite the best efforts of ACDA and SIPRI. Moreover, the content of 'defence expenditure' and its valuation often appear ambiguous. In terms of the former, the magnitude of the defence effort may vary by alarming proportions depending on the definition adopted.[1] As for the latter, serious problems in valuation can be occasioned by inflation and exchange rate effects, not to speak of distortions caused by the difficulty of pricing non-marketable commodities. As if this was not enough, one dollar of military expenditure in one country does not necessarily deliver the same amount of capability as a dollar spent in another country owing to widely differing technological, production and manpower endowments. All told, then, any utilisation of aggregate defence expenditure data, whether from international agencies or directly from national accounts, must be treated with the utmost caution.

Caution notwithstanding, some aggregate trends are immensely telling. Most salutary is the sustained increase in world military spending despite sluggish and uneven growth in the global economy. In fact, military expenditure now approaches $1,000,000,000,000 and, what is more, has been subject to accelerating growth. Thus, while the average annual real growth of world military expenditure between 1975 and 1980 registered 1.5 per cent, the figure had leapt to 3.2 per cent in the succeeding quinquennium--and did so despite a rate of increase in world GDP of only 2.4 per cent.[2] These aggregate figures, however, obscure considerable geographical variations. For the developing countries as a whole, GDP grew at an annual rate of 1.8 per cent whereas military expenditure expanded at 3.1 per cent. For countries with per capita incomes of less than $440, real military spending has grown at a rate of four per cent per annum since 1976. Yet most weight must be attached to the activities of NATO and the Warsaw Pact which together account for 73.5 per cent of global outlays in roughly equal proportion. The USA and USSR are dominant within each bloc, with the former, for example, arrogating more than 60 per cent of total NATO outlays. It and its alliance partners have been committed since 1978 to the Long Term Defence Programme, an augmentation strategy requiring each member to accede to annual real defence expenditure increases of approximately three per cent. In the event, this target was not consistently adhered to, although defence budgets increased everywhere. A principal outcome was the increase in the share of defence outlays earmarked for equipment (Table 1.1).

In general, global patterns of defence equipment purchases have been on the rise in conjunction with changes in the form of defence production, the emergence of new defence industries--in both geographical and technological senses of the term--and the restructuring of defence procurement policy. The processes underpinning these changes are pervasive, with the most critical being the incessant drive on the part of the advanced-industrial countries (AICs) to generate and maintain technological advantages and its corresponding response, that is to say, the desire of the newly-industrialising countries (NICs) to acquire as much defence production capability as possible. Essentially, the post-World War 2 division of labour in defence production, marked by the overwhelming dominance of the USA and USSR, is giving way to a more stratified division among a greater number of producers.[3] Position in the global defence industry hierarchy is a function of the size of the national economy, that enabling agent providing the resource base necessary to initiate defence production and secure production economies, as well as the level of development of the nation's technology. The two are to a certain degree interdependent and both are necessary if across-the-board capabilities are to be established and retained. Thus, while a country like China possesses size sufficient to indigenously provide comprehensive capabilities, technological constraints necessitate the import of

Table 1.1 : Equipment Share of Total Defence Outlays

	1973-1977	1978/9	1979/80	1980/1	1981/2	1982/3	1983/4
Belgium	10.0	13.9	13.1	14.4	14.0	13.6	14.9
Canada	7.3	10.0	13.8	15.4	15.9	17.4	18.8
Denmark	19.6	16.4	16.3	18.1	17.5	16.8	17.3
W. Germany	12.3	13.0	13.7	14.8	17.3	17.3	17.6
Greece	18.3	18.1	17.7	18.8	20.5	17.0	18.9
Italy	14.5	16.2	15.1	17.5	17.3	13.2	18.5
Netherlands	16.0	18.3	20.2	18.0	18.8	20.4	22.0
Norway	14.1	18.3	19.5	19.3	19.0	19.5	18.3
Turkey	21.0	18.5	9.1	4.7	9.4	10.8	10.1
UK	20.0	23.0	23.2	25.2	26.5	25.4	28.2
USA	18.3	20.0	19.5	20.3	21.3	23.9	26.1

Source: NATO, Facts and Figures 1984.

advanced systems to upgrade weapon performance. By way of contrast, countries such as the UK and France occupy an intermediate position, producing much but not all of the technological spectrum of weapons systems. The UK's attempt, for example, to develop domestically an AEW aircraft is an apt illustration of technical and financial constraints collectively impinging on the scope of the defence-industrial base. After nearly £1 billion of Nimrod programme development expenditure, continuing problems with the mission system avionics compelled the MoD to forsake the home product in favour of the E-3A AWACS supplied by US contractors (a course subsequently adopted by France).

As with other products, international competitiveness in defence production rests, in part, on principles of comparative advantage and functional specialisation. In turn, these economic truisms rely on a crucial 'software' input; namely, the capacity to design and develop effective weapons systems. Of

vital importance in this regard is the willingness
to devote resources to military R & D. Table 1.2
offers a rough indication of the design capacity of
defence industries in the AICs simply on the premise
that more expenditures channelled into R & D should
be forthcoming with deeper weapons system expertise.
The USA and USSR together accounted for some 80 per
cent of the $80 billion (current prices) devoted to
global military R & D in 1985. The addition of the
UK, France, China and West Germany places a select
group of countries in the position of accounting for
fully 90 per cent of world military R & D expenditures. With technology given pride of place in the
defence industries of AICs, the importance of R & D
is obvious. In both the USA and UK, for instance,
the sustained increases in defence spending have
especially favoured R & D allocations. For example,
the Reagan build-up witnessed defence R & D outlays
escalate from a 1980 level of $7.7 billion (in 1972
dollars) to $11.4 billion in 1984, a real increase
of 48 per cent.[4] Among the more important technological ventures undertaken during this period are
the Very High Speed Integrated Circuit programme and
the Strategic Defence Initiative (SDI). For its
part, R & D expenditures in the UK climbed from £1.6
billion in 1980-1 to £2.1 billion in 1984-5, an
almost 30 per cent increase.[5]

The implication of the concentration of defence
R & D outlays is that it clearly places the leading
edge of defence technology in the hands of only a
few producers. Control over technological resources
is a key factor in determining which countries are
likely to be in a position to occupy the early
stages of the defence product cycle. This technological leadership accorded to the AICs has far-
reaching consequences in setting the technological
agenda for future arms production and thereby per-
petuating the existing hierarchy of defence industries. The most important area of emerging technology is defence electronics which, in the words of
the US Defense Science Board, 'is the foundation
upon which much of our defense strategy and capabilities are built', and is the basis of enhancing performance and survivability of weapons systems.[6] Consistent with such thinking is the increasing share
of defence spending devoted to the electronics sector. In the USA, the electronics share of the R & D
budget is scheduled to grow from 49 per cent in 1987
to 54 per cent in 1996 while its share of the procurement budget is to rise from 35 to 40 per cent
and amount to a staggering $51 billion. The increas-
ing importance of electronics is attributable to

Table 1.2 : Military R & D Expenditure

	US$ millions, 1980 prices and exchange rates				
	1976	1978	1980	1982	1984
USA	15785.4	16206.8	15766.5	19386.4	23798.0*
UK	2959.7*	3181.8*	3718.7*	3354.0	3695.2
France	1989.7	2235.2	2685.8	3116.5	3111.0*
W. Germany	959.1	1047.0	951.8	809.6	899.9
Sweden	288.7[2]	323.0	228.6	246.4	360.0*
Italy	56.1	59.5	48.7	121.4	260.0*
Japan	102.0	116.6	125.6	144.2*	169.6*
Australia	125.6[3]	na	110.2	102.3	103.1*
Canada	83.5	84.3	83.2	87.9	na

Notes: 1. Figures marked by asterisks are either provisional figures or
SIPRI estimates.
2. Fiscal year 1975-76.
3. Fiscal year 1976-77.

Source: Compiled from SIPRI Yearbook 1986, p. 301.

expanded use of lasers, infra-red devices, fibre
optics and electro-optics for surveillance, target
location and designation, communications and guidance as well as the wider use of millimetre wave
hardware for secure communications and weapons guidance.[7] Particularly critical are guidance and surveillance systems and, significantly, production
capability in these areas is almost exclusively the
preserve of a few AIC defence industries.[8] The concentration of defence R & D is additionally meaningful in that it points to the dependence of smaller
producers on the transfer of technology. For these
producers, entry into defence markets is generally
predicated on their ability to select niches where
price competition is relatively more important than
the preoccupation of the major producers; namely,
performance criteria. By the same token, market
access for those smaller defence industries is contingent on their obtaining quality sub-systems from

5

the AICs for the products geared to market niches.
For example, Brazil's successes in the global arms
trade would have been impossible without access to
foreign sub-systems. EMBRAER's best-selling Tucano
military training aircraft utilises either Garrett
or P & WC engines from the USA and Canada respectively, Collins avionics from the USA, and Martin
Baker ejection seats from the UK.

PROCUREMENT POLICIES AND DEFENCE INDUSTRIES

To varying degrees, most AICs have pursued policies
designed to attain a degree of capability and independence in defence manufacture, be it for so-called
'strategic' reasons or simply to remain abreast of
potentially important technological developments.
Yet, all weapons programmes have been confronted
with the seemingly inescapable problem of rapidly
escalating development and production costs as newer
systems replace their outdated predecessors. As a
result, the objective of fostering and maintaining
an indigenous defence-industrial base has invariably
been circumscribed by the prohibitive costs of successive weapons systems. In a nutshell, defence
planners are faced with the central problem of balancing available resources, military-industrial
requirements, and desired force levels and capabilities.[9] Without doubt, the primary cause of such
weapon cost escalation is technological change by
virtue of the fact that each successive system
incorporates new technologies in the quest for
improved performance. To dramatize the point, it is
only necessary to acknowledge precipitous intergenerational declines in the numbers of weapons produced. Thus, while total US defence budgets (as
measured in current dollars) were roughly similar at
the height of World War 2 and in 1974, the former
could buy 50,000 tactical aircraft, 20,000 MBTs and
80,000 artillery pieces whereas the latter had to
settle for 600 aircraft, 450 tanks and no artillery
pieces at all.[10] Evidently, weapons systems--regard-
less of type--have been subject to sharp increases
in replacement costs in modern times and the process
is not relenting. A contemporary example from many
is the GM Hughes AMRAAM, an air-to-air missile which
is estimated to be about 40 per cent more expensive
than the Raytheon Sparrow AAM it is designed to
replace.[11] Yet, even this cost escalation may be an
understatement. Critics of the AMRAAM demur with
released figures, claiming that they are based on
impending production improvements: rather than a

'modest' 40 per cent surcharge on the $171,000 cost of a Sparrow, the critics bruit about a figure of $750,000 per AMRAAM. A multiplicity of factors underscore the problem of weapon system cost growth of the kind afflicting missile programmes. However, two fundamental causes deserve singling out: the nature of the weapon succession process and the national structure of defence manufacture. While Chapter 3 will give them full measure, these notions warrant some preliminary remarks at this juncture.

The Weapon Succession Process

The process by which one weapon system is followed by another encapsulates, quite simply, the process of military-industrial change.[12] The weapon succession process involves the interaction between demand and supply as they relate through the life-cycle of military technology. The crucial issue at hand is the means whereby the appropriate quantities of resources devoted to defence are reconciled with a set of security needs in the absence of war. The key feature characterising the demand side is the conservatism of military institutions and bureaucracies: a corollary both of organisational inertia and the natural wariness of such bodies to the prospect of social and political disruption. Paradoxically, the military is ready to embrace change as expressed through new weapons systems: the weapons, after all, are but tools to consolidate the status quo. The supply side, meanwhile, is invested with two basic considerations; namely, the types of suppliers involved in the design, development and production of weapons systems and, secondly, the nature of contracting procedures. Dominating the former are specialised defence firms that must forever compete for defence contracts so as to maintain capacity usage and, of course, profits. Their very specialisation often inhibits these firms from diversifying into commercial markets. Theirs is the condition of relying on a succession of weapons programmes in order to avoid corporate demise. In that light, they stress their technological assets and ability to provide 'superior' weapons as determined by performance criteria. The second consideration follows from the glaring fact that defence firms are also important entities within the economy at large, not only in a sectoral sense of occupying key industries and a spatial sense of affecting particular regions and localities, but in a political sense of having influential interest groups associated with them. In

7

short, a combination of circumstances works to further strengthen the need to ensure their survival (a theme incident to Chapters 2 and 3). Indeed, there is a high degree of interdependence between industrial and regional policies and defence procurement policy: all conspiring to perpetuate defence firms by means of sustaining contracts for military hardware. Consequently, the weapon succession process is institutionalised through contract procedures which purport to secure the survival of suppliers while fulfilling the established needs of military bureaucracies. Since competition for weapons supply revolves round the 'superiority' characteristics demanded by those bureaucracies, challenging specifications push designs to the cutting edge of technological capabilities while, as aforementioned, accelerating unit costs.

National Production Structures

Pressures on defence budgets have not been moderated by the national production frameworks through which defence industries operate. Nowhere is this more apparent than in the workings of NATO. Although NATO is a supranational military alliance, responsibility for weapons R & D and production remains vested with national structures. Furthermore, the alliance consists of a group of nations which actively and aggressively compete with each other across a range of political and economic fronts. This rivalry is manifested in a number of ways, not least of which is military-industrial production. Sadly, the situation shall remain inviolable so long as decisions concerning weapons manufacture are inextricably linked to ideas of national sovereignty and prestige as well as the practicalities of industrial and regional policy. The contentious trade dispute surrounding large civil airliners (Airbus) is ample evidence of the tight interdependence of these policy areas, as the Europeans adamantly argue that DoD contracts are direct subsidies to Boeing and McDonnell Douglas (MD). Similarly, SDI is also regarded as a technological challenge by America's allies. To a large extent, then, defence procurement decisions are informed first by national considerations and only secondarily by greater NATO needs. As a result, the bloc members must rub along with extensive duplication of defence R & D and production capacity, smaller production runs, unrealised scale and learning economies, and a plethora of weapons types fulfilling equivalent military functions. In short,

the defence market within NATO is highly fragmented.
Evidence of the cost of fragmentation can be calculated. It has been averred, for instance, that NATO
standardisation could yield weapons unit cost savings in the order of at least 20-30 per cent.[13]

Restructuring Procurement and Supply

Both state procurement policy and production systems
must be restructured if a semblance of fiscal control is to return to the weapon succession process.
Broadly speaking, procurement reforms are aimed,
first, at coming to grips with technological and
financial constraints associated with advanced weapons development and manufacture and, secondly, at
improving the efficiency of transforming outlays
into desired capabilities. Reforms have been per-
formed along two major fronts. On the one hand,
emphasis has been placed on better management and
cost effectiveness. For example, the DoD replaced
annual funding arrangements with multi-year allocations when it was given to understand that contractors would be better able to order long-lead items
and achieve scale economies under the new framework.
On the other hand, attention has been directed to
overcoming the inefficiency of defence firms. Inefficiency derives from four sources: monopolistic and
oligopolistic market structures, barriers to entry
and exit, non-competitive defence contracting practices, and government regulation of defence-industry
profits.[14] The characteristics are occasioned by the
protection afforded defence firms from the rigours
of the market. The dismantling of the protective
cocoon has commenced in the AICs with the forced
introduction of greater competition into contracting
practices. In the USA, for example, 'second-sourcing' of fighter engines, missiles, electronics and,
increasingly, warships is now the rule in place of
the old practice whereby a single prime contractor
retained responsibility for the entire production
run of a weapon system. Along the same lines, competition in defence markets is now the hallmark of
UK policy. Whereas in 1979/80 only 14 per cent (by
value) of MoD contracts were priced through competitive bidding, the proportion had risen to 38 per
cent by 1985/6. Savings accruing from these changes
have been far from inconsequential: for example,
winning tenders for training missiles and torpedo
warheads were 50 and 16 per cent less than the previous non-competitive awards.[15]
  In addition to reforming procurement practices,

governments have intervened--in the hunt for
improved efficiency--to restructure defence industries. The current UK Government has favoured privatisation of state enterprises as its efficiency-pro-
motion measure and RO, BS and BAe have been
indelibly transformed as a result. In the USA of
strictly private defence firms, the DoD has instituted programmes designed to improve company productivity (and, therein, reduce weapons costs). Thus,
the Industrial Modernisation Incentives Programme
was constituted to address two problems.[16] In the
first place was the inherent inability of cost-plus
contracts to contain cost inflation while, secondly,
was the dismay felt by contractors at the possibility of programme cancellations, cut-backs or slow-
downs. To foil these drawbacks, it was decided to
offer investment protection guarantees and incentive
payments based on cost reductions and avoidances.
One outcome has been a three per cent diminution in
the airframe costs of the F-16 fighter.

INTERNATIONAL DEFENCE PRODUCTION

A more profound consequence of changing procurement
policies is the 'internationalisation' of defence
manufacture. This has occurred in the form of international collaboration in the design, development
and production of weapons systems. Unlike the 'internationalisation' of capital in many civilian sectors, its guise in the defence industries has neither relied on direct investment nor has it been
marked by the search for low-wage locations. In
other words, establishing branch plants as part of
world-wide sourcing strategies for re-export has not
been particularly pronounced among defence firms.
Instead, the motive espoused by the firms and abet-
ted by their governments is to share overhead costs
among international partners while retaining, as far
as is feasible, national production structures. The
UK experience in combat aircraft manufacture underscores the inducements promoting the adoption of
such an approach. Specifically, the Plowden Committee candidly argued in the mid-1960s 'that there may
still be a case on defence grounds for some domestic
capacity to produce military aircraft and guided
weapons, but there is no longer a case for providing
all our defence requirements at home'.[17] The Committee went on to recommend that 'Britain must turn to
collaboration with other countries as the means for
improving the relationship between sales and development and initial production costs for aircraft and

guided weapons', and advocated close ties with European countries confronted with comparable problems. Systems such as the SEPECAT Jaguar, Aérospatiale/Westland Puma, Gazelle and Lynx helicopters, and the Panavia Tornado were concrete manifestations of the official reception of Plowden's message. Nor has the enthusiasm for collaborative ventures waned: as of May 1987 the UK was involved in 37 international weapons systems.

A second form of 'internationalisation' being pursued is that resulting from the push to export weapons so as to capture production economies, maintain capacity and offset R & D costs. The pressures to export are exacerbated when domestic and collaborative contracts are running down. At that stage governments are compelled either to respond with short-term palliatives, as in the case of West Germany's decision to order an extra 150 Leopard II MBTs to boost the ailing tank industry operating at just 60 per cent of capacity, or to foster defence exports.[18] A number of policies have been adopted to encourage defence firms to export. New arrivals on the scene (e.g. China and Brazil) as well as more established producers (e.g. France and Italy) have pursued relatively 'open door' strategies of not discriminating, on political grounds, against potential clients. Governments have also backed the firms through export sales support, as is the case in the USA with Foreign Military Sales credits and in the UK through the Defence Export Services Organisation. In terms of actual weapon design and development, export market dependence implies that governments and firms are no longer at liberty to exclusively tailor systems to domestic needs. Rather, effective export development strategies require weapons systems couched to fit a multiplicity of foreign needs as well as those of the home government. Dassault-Breguet's Military Sales Manager, Paul Jaillard, argues that to justify the enormous development costs of modern warplanes, programmes must be flexible and sensitive to a range of markets. In his words, 'If we make an aircraft suitable only for Europe, that will not be satisfactory'.[19] In that light, 163 of the first 276 firm orders for the Mirage 2000 fighter were booked by export customers.[20]

Confounding the export option, however, has been two factors; to wit, the tendency whereby the transfer of weapons production technology has gained emphasis relative to the outright transfer of military end-products and, secondly, the recent depressed global market for defence items made worse

11

by the emergence of new NIC entrants into the market. The two, of course, are interrelated since the emphasis on weapons production technology transfer gives rise to new capacity which, ironically, limits future market inroads for existing producers. Fragmentation of the arms market is an inevitable consequence, and it leads to defence firms seeking solace as niche players. This state of affairs has come about as a sequel to the straitened global arms market following in the footsteps of economic recession. The remaining customers have a strengthened bargaining position and are using that leverage to insist on industrial offset agreements, often entailing the erection of 'instant' defence industries or the sustenance of the existing defence-industrial structure. The UK's purchase of AWACS fits readily into the latter mould. In March 1987, the MoD agreed to purchase six Boeing E-3A aircraft at an estimated price of £860 million.[21] Boeing and its system partners assented, as a condition of the sale, to industrial offsets to UK firms worth 130 per cent of the contract value. These were to be spun out over a period of eight years commencing with the time of purchase and were subject to penalties in the event of non-performance. Also, Boeing is bound to a technology clause which stipulates that the offsets arranged by it must be of a technical standard commensurate with that represented by the AWACS. Offsets are both programme related and independent of AWACS production. The former consist for the most part of installation and check-out work based on technology provided by Boeing and upwards of 100 UK firms have visited Seattle for briefings on the contracts in the offing. Work falling into this category includes systems installation, flight testing and depot-level life-cycle support. The work unrelated to AWACS, meanwhile, embraces such contracts as that between Boeing and Racal aimed at establishing a tactical radio facility in Saudi Arabia (as part of the US company's 'Peace Shield' offset programme with that country), the Boeing/Perkins deal engineered to supply diesel powerplants for the hard mobile launchers of the US small ICBM project, and the arrangement whereby AWACS-systems supplier, Westinghouse, sponsors radar development at Plessey. The above is merely one, albeit highly visible, case of the systematic use of international market leverage on the part of purchasing governments to fuse defence procurement policy and strategies of economic development. Other examples are commonplace. Indonesia, for instance, has linked its defence procurement policy to its goal of developing

a national aerospace industry.[22] Its recent $337
million purchase of GD F-16 fighters carried with it
an obligation for the American seller to market
Indonesia's own CN-235 transport aircraft.[23] The
CN-235 is a joint venture of Indonesian (IPTN) and
Spanish (CASA) state firms and forms an integral
component of the efforts undertaken by both countries to enter new export markets in an area of
advanced manufacture. Spain is also tying defence
procurement decisions based on its market leverage
to industrial policy objectives. To this end, the $3
billion purchase of F-18 fighters from MD is linked
to industrial offsets centred on the kind of technology transfer consistent with the country's development of aeronautical production capabilities.
Through the F-18 project, Spain is gaining access to
carbon fibre epoxy bonding, numerical-control
machining, aluminium honeycomb core carving and ion
vapour deposition techniques.[24] Purchasing countries
have thus been able to promote import-substitution-
industrialisation at the same time as they have augmented defence industry. These efforts have served
as the rationale for NIC entry into defence production, a topic dealt with in detail in Chapters 4 and
5.

All told, procurement patterns are shifting
under the impetus of technological change and the
'internationalisation' of defence industries. With
few exceptions, the concept of the sovereign, or
fully independent, national defence industry is more
a legacy of wishful hankerings for nationalist glory
than a viable policy option in today's world. To be
sure, governments continue to subscribe to the ideal
of an indigenous defence-industrial establishment
but, in truth, this aim has to be tempered by emerging realities of weapons production: realities which
increasingly take on the mantle of international
links between defence firms. The evolution of this
process is scarcely governed by the unfolding of
pure principles of comparative advantage: rather, it
is simply of a piece with the intersecting of mili-
tary/political desires on the one hand, and econom-
ic/technical ones on the other. These issues form
the backcloth to an understanding of the constituent
units, the firms, of defence industries and, as
such, have been aired before examination of those
constituents becomes admissible. So far, we have
referred to defence industries without clarifying
from what sectors they are composed. It is to this
subject that we now turn.

KEY DEFENCE INDUSTRIES

Defence industry, as Kennedy remarks, is an ambiguous term without definitional substance.[25] At one extreme, it embraces industrial sectors that unequivocally manufacture military goods--for example, the makers of artillery and submarines--but, by and large, it includes sectors that produce civilian goods too. Thus, builders of frigates may also engage in passenger ferry construction while warplane makers may be equally conversant with civil jetliner production. Individual defence industries, then, are less distinct and unique sectors geared to defence markets and more general manufacturing sectors producing for both military and civil needs are distinguished from each other by their specific technological and material input-output requirements. Their designation as defence industries rests on the destination of the bulk of their output: should most of it be earmarked for defence markets, the industry is presumed to be a defence industry. However, even such a simple definition as this rule of thumb is not without its ambiguities. For example, the orientation of demand may vary through time. By way of illustration, the percentage value of naval work relative to total UK shipbuilding output between 1957 and 1964 registered the following annual values: 9.9, 13.7, 16.0, 13.4, 11.8, 20.4, 19.5 and 23.6.[26] In the space of eight years the proportion increased more than twofold. Again, British shipbuilding employed 120,670 in naval work in February 1945 and 98,153 in merchant work: by June 1948 only 3,174 were occupied in naval work as against 170,486 working in merchant shipbuilding.[27] Obviously, the market bias of an industry is affected by emergency defence requirements. Furthermore, the importance of defence demand to an industry will vary by country, reflecting the particular structure of a nation's armed forces. In the 1970s, for instance, domestic military output amounted to 51.6 per cent of total domestic final output for the UK shipbuilding industry, 3.0 per cent for the automotive industry and 12.9 per cent for instrument engineering. The equivalent American figures were 50.4, 1.5 and 9.8; a small but meaningful cross-national disparity.[28] And last but not least, the importance of defence demand will differ according to enterprise. Dependence on the military varies quite markedly even in the US aerospace industry, the classic core of the military-industrial complex. Reliance on DoD and NASA contracts in 1979 ranged from 19.9 per cent of total sales in the

case of Boeing to 91.8 per cent in the case of Grumman.[29]

Examination of the industries influenced by
superpower defence demand is one means of empiri-
cally determining the scope of defence industry. It
is evident from Table 1.3 that some manufacturing
industries are well and truly defence industries in
so far as the bulk of their output is destined for
the US DoD. The ordnance, complete guided missiles
and radio & communications equipment sectors all
conform to this category; as indeed does the shipbuilding & repair industry. For its part, the aircraft industry adopts an intermediate position with
about two-fifths of its output steered to the military. The other defence suppliers are much less
reliant on defence demand, although the proviso
remains that some firms within these latter industries may be much more oriented to serving defence
customers. All in all, ten US manufacturing industries accounted for 70 per cent of DoD industrial
purchases. The leading five--aircraft, communications equipment, missiles, ordnance and shipbuild-
ing--alone absorbed 55 per cent of the value of purchases.[30] While comparable Soviet information is
absent, the USSR does, nevertheless, identify similar activities within the ambit of its defence production ministries. Consequently, the Ministry of
General Machine Building is tasked with the production of rockets and space equipment, Machine Building is held responsible for munitions (ordnance),
the Shipbuilding Industry and Aviation Industry ministries are self-explanatory; the remit of Defence
Industry is conventional arms, Radio Industry deals
with radios, Electronic Industry with radars, while
the Communications and Equipment Industry ministry
provides all other communications apparatus. Not to
be overlooked is the Ministry of Medium Machine
Building which supplies nuclear equipment for
defence purposes.[31] In like fashion, China desig-
nates various state organisations as military ventures: the Ministries of Nuclear Industry (nuclear
weapons), Aviation Industry (aircraft), Electronics
Industry, Ordnance Industry and Space Industry (ballistic missiles) as well as the China State Shipbuilding Corporation (CSSC).[32] It is evident, therefore, that defence industries are constituted, in
varying degrees, from a bundle of manufacturing
industries drawn from the mechanical and electronic
engineering sectors. The operative phrase, however,
is 'in varying degrees'. In the final analysis it is
the defence-orientation of the particular enterprise
which determines the extent of defence industry.

Table 1.3 : Industries Serving the DoD, 1979

Industry	DoD demand ($ million)	% of industry total
Aircraft	11,754.2	39.6
Radio & communications equipment	10,795.9	51.8
Complete guided missiles	4,277.0	58.2
Ordnance	3,747.6	65.0
Shipbuilding & repair	3,424.5	50.4
Motor vehicles	1,744.5	1.5
Industrial chemicals	1,269.5	16.2
Petroleum refining	1,106.3	3.4
Computers & peripheral equipment	1,046.7	4.6
Optical & photographic equipment	802.7	5.0
Scientific & controlling instruments	736.9	9.8

Source: Derived from R.W. DeGrasse, 'Military Spending and Jobs', Challenge, July-August 1983, p. 10.

Defence-orientation is defined as the degree to which any firm is reliant on defence contracts for its continued viability and, as intimated, it varies from country to country and from time to time.[33] Where defence-orientation is clearly in the ascendant, the enterprise is said to be defence-dependent.

During 1976/7 only five UK firms received contracts worth more than £100 million from the MoD and they were presumably defence-oriented in consequence.[34] Three of the five belonged to the aerospace industry (i.e. the BAe predecessor companies of British Aircraft Corporation and Hawker Siddeley Group, along with Rolls-Royce), one was an electronics firm (GEC) and the fifth was the then state-owned Royal Ordnance Factories (now RO). A true indication of defence-dependency can be got for US firms when the value of their prime contract awards from the DoD are set against their total revenues.

This ratio is displayed in Table 1.4 for firms
deemed vital to the American defence effort. Since
the ratios are given for each firm for a run of five
years, it is interesting to note the mutability of a
firm's defence orientation. Some enterprises record
consistently high ratios and hence, defence dependence: the aerospace giants of General Dynamics
(GD), Lockheed and MD are especially prominent on
this score. Others--particularly the electronics
specialists (e.g.   GTE, Honeywell,   Singer and
TRW)--are much less dependent on Pentagon demand.
Overall, though,   the defence-orientation of this
clutch of firms was accentuated over the 1978-82
period as a direct result of the transition from the
Carter regime to that of Reagan with the latter's
penchant for enlarged defence budgets.

Defence Dependence over the Long Run

The defence orientation of a firm or industry is
obviously very sensitive to changes in the aggregate
size of defence budgets.   Wartime or rearmament
budgets not only act to augment the defence workload
of existing arms contractors but they conspire to
enlarge the scope of defence industry, drawing into
its compass sectors which generally would be preoc-
cupied manufacturing for civil markets. A traditional industrial agglomeration of the likes of Birmingham and the Black Country provides a cogent
example. The effect of World War 1 was to expand
the drop-forgings industry severalfold as this was a
basic supplier of shell and ordnance industry. However, the metals industry was stimulated in more
unconventional ways: the phoenix-like growth in the
demand for aeroengines led to a great expansion in
the aluminium-castings business while the moribund
pig-iron industry witnessed a dramatic rise in its
fortunes.[35] In short, wartime demand was (and, in
general, is) sufficient to lift declining sectors
out of depression and, simultaneously, create new
businesses altogether. By the same token, sharply
curtailed defence budgets not only provoke cutbacks
and the return of 'conscripted' industry to civil
tasks, but they can result in the demise of defence-
oriented enterprises unable to adjust to reduced
military procurement. In the case of Birmingham,
many firms in the automotive industry found reverting to postwar car production particularly troublesome as a result of their inheritance of special-
purpose munitions plant which could not readily be
turned over to civil lines. A similar situation

Table 1.4 : Defence Dependency of US Firms

| Firm | Ratio of DoD prime contracts to total revenues | | | | |
	1978	1979	1980	1981	1982
Avco[1]	.08	.08	.15	.23	.27
Boeing	.27	.18	.24	.27	.35
GTE	.03	.03	.03	.04	.05
GD	1.30	.86	.76	.67	.96
GE	.09	.09	.09	.11	.14
General Tire & Rubber[2]	.08	.11	.17	.15	.30
Honeywell	.15	.16	.14	.16	.22
Litton	.43	.20	.15	.28	.27
Lockheed	.64	.44	.38	.43	.62
Martin Marietta	.31	.25	.31	.39	.47
MD	.69	.61	.54	.60	.77
Raytheon	.35	.29	.35	.32	.41
RI	.17	.11	.14	.16	.36
Singer	.11	.13	.16	.21	.22
TRW	.09	.10	.10	.10	.17
Tenneco[3]	.05	.10	.12	.07	.05
UTC	.38	.28	.25	.28	.31
Westinghouse Electric	.08	.09	.11	.12	.15

Notes: 1. Now part of Textron
2. Through its Aerojet-General subsidiary
3. Through its Newport News Shipbuilding subsidiary

Source: Derived from K.A. Bertsch and L.S. Shaw, The Nuclear Weapons Industry
(Investor Responsibility Research Center, Washington, DC, 1984).

pertained to Japan at about this time. Many of the
largest factories in Osaka and Tokyo had become
reliant for their very survival on defence contracts
by 1920. The firms in question took in the cream of
Japan's modern industry, including metal refining
(e.g. Sumitomo Sheet Copper Manufactory and the
Osaka Zinc Company), castings (e.g. the Tobata Casting Company and the Ota Casting Mill), wire making
(e.g. the firms of Yokohama, Sumitomo, Fujikura and
Nippon Wire), rubber manufacture (e.g. the Meiji,
Mitado and Nippon Rubber companies) and, of course,
shipbuilding (e.g. Kawasaki Dockyards).[36] They too

had difficulties adjusting to reduced defence budgets in the 1920s.
The fact that defence budgets are subject to
wide swings is readily demonstrated. In the last
half century and more, for instance, defence spending in the USA as a percentage of GNP has ranged
from a low of 0.88 in 1930 to a high of 41.52 in
1944. Throughout the 1930s, in fact, budgets tended
to stabilise round the 1.25 mark; a continuum rudely
interrupted by war and budgets climbing from 10.97
in 1941 to the peak 1944 figure. The return of peace
saw the 1945 level of 34.62 plunge to a postwar low
of 3.87 in 1947. Korean War demands pushed the figures back up to over ten per cent and the subsequent
Cold War and Vietnam episodes kept the budgets in
the seven to nine per cent band through to the
beginning of the 1970s (whereupon they progressively
declined to about five per cent by the end of the
decade).[37] Converted into constant (1957) dollars,
defence outlays were ticking over at an annual rate
of two to three billion in the 1930s, but zoomed to
an atmospheric $153.1 billion in 1944 and, in short
order, dropped to $11.8 billion in 1947 before
steadying into a $40-60 billion range for the suc-
ceeding decades. In practical terms, boosted
defence budgets translate into expanded order books
for defence industry whereas budgetary restraint not
only diminishes the size of the order books, but can
mean outright cancellations of orders. In so far as
the former is concerned, the US situation in World
War 2 is instructive. Starting from a 1939 index of
production set at 100, capacity in the aircraft
industry was expanded in short order to attain a
production peak of 2,805. Other impressive performances were achieved by the ordnance (2,033), shipbuilding (1,710), locomotive manufacturing (770),
aluminium (561), industrial chemicals (337), rubber
products (206) and steel (202) industries. All of
these industries subsequently suffered readjustment
pains, especially the ones that had expanded the
most.

At the enterprise level, such oscillations in
defence spending can play havoc with capacity,
resulting in full utilisation in times of rising
budgets and crippling under-utilisation during times
of retrenchment. Figures 1.1 and 1.2 display longitudinal profiles of a century's output of naval tonnage undertaken by four prominent warship constructors.[38] The four are selected because they are major
elements in the shipbuilding industries of the UK
(Vickers), the USA (Newport News), France (Arsenal

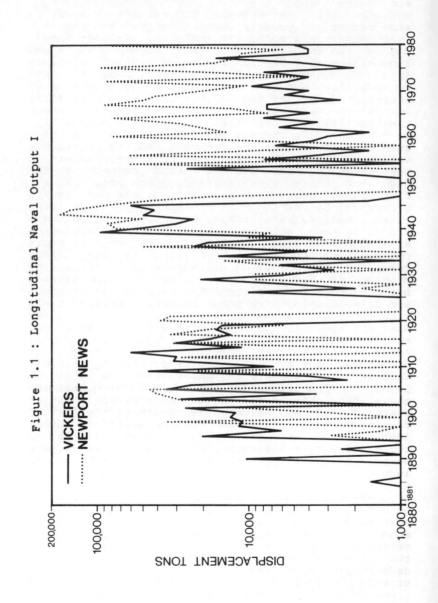

Figure 1.1 : Longitudinal Naval Output I

de Brest) and Japan (Mitsubishi). While these
enterprises shall receive more detailed treatment
elsewhere, it is necessary to know that three of
them represent private capital while the fourth--
Brest--is a state undertaking.[39] Government ownership worked in favour of Brest for much of the time:
it consistently launched warships when the private
yards were confronted with work shortages (the main
voids in its launching records derive from the dis-
location caused by German occupation in World War
2). The private yards did not enter naval construction until later (1886 for the UK enterprise, 1895
for Newport News and 1906 for Mitsubishi), a consequence of the predilection of admiralties to reserve
warship building for their own navy yards. This
preference was only overcome when demand exceeded
available state capacity and could only be met
through the involvement of private yards. As can be
elicited from the figures, output in all four countries steadily expanded from 1895 until World War 1
and the enterprises in question were major beneficiaries of this rising tempo of demand. However, output patterns diverge thereafter: the UK and France
reduced aggregate output as onus was switched to the
production of smaller escort vessels and submarines;
but, as befits their status as latecomers, the USA
and Japan continued to boost their main battlefleets
right up until the Washington Treaty of 1921. International agreements on fleet sizes moderated output
at all four shipbuilders throughout the bulk of the
Interwar period until preparations for renewed war
drove tonnage figures upwards in the late-1930s. The
immediate postwar era was marked by low activity
levels in the shipyards and it was not until the
early-1950s that orders began to pick up again in
response to Korean War demands. Subsequently, all of
the yards have maintained steady levels of output
(more intermittent in the case of Brest), although
the absolute size of the output issuing from Newport
News has overshadowed the other producers. In the
long run, then, naval shipbuilders--and defence suppliers as a whole--experience great variations in
production levels in response to shifting patterns
of demand. These variations differ in detail since
each country's policies, industrial structure and
enterprises are specific to it, but all countries
respond to the same basic stimuli if only because
they operate in an interdependent world where
defence considerations take stock of the behaviour
of other nations.

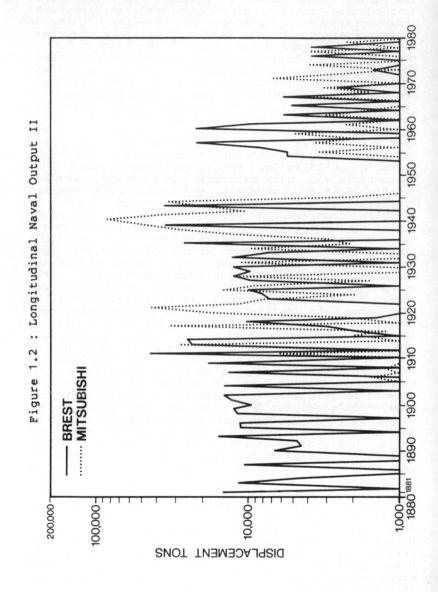

Figure 1.2 : Longitudinal Naval Output II

THE NATURE OF DEMAND

A brief sally into the output performance of leading
defence firms over the years calls for an elabora-
tion of the nature of demand for their products. Put
otherwise, the issue reduces to one of determining
whether any systematic pattern is discernible in the
cycle of defence procurement. As has been made abundantly clear, severe fluctuations characterise the
temporal profile of the output of the naval manufacturers. Can such swings in production be made com-
patible with a demand cycle and, if so, can such a
cycle be generalised as coming to bear on all
defence firms and not just those fulfilling naval
orders? As it happens, defence production rarely
conforms to the fluctuations of the business cycle;
rather, it subscribes to the dictates of international tension. A descriptive model outlining how
rivalry--both real and latent--between the great
powers (and, by extension, the regional powers as
well) amounts to a 'wave-cycle' in arms ordering and
production appears to be much more appropriate to
defence demand than any business cycle rationale.
In brief, this model is composed of four basic constituents which represent periodic production
phases. What might be termed normal production--
peacetime equilibrium--represents replacement
demand, that is, splicing into production new items
as and when old equipment reaches the end of its
useful life. Disruptions in this equilibrium occur
whenever policymakers perceive of a threat which can
only be confronted by boosted defence forces (i.e.
the rearmamental instability period). Expansion may
entail industrial mobilisation and, should it tran-
spire that large-scale hostilities come about, then
mass-production of defence equipment is the overriding concern (and represents the wartime equilibrium
period). A transition period occurs with the cessa-
tion of hostilities (known as demobilisational
instability) and this takes in the often painful
enforced conversion of industry to civilian production. Thus, the wave-cycle model of procurement has
been put forward to account for precisely the pattern of surges of production, rapid cutbacks and
steady ordering of the replacement kind that could
be inferred from the longitudinal output records of
the four previously mentioned naval shipbuilders.[40]
However, the wave-cycle idea, in turn, derives from
examination of military aircraft production trends,
especially as evidenced by the UK record but also
encompassing US experience.[41] Accordingly, the whole
idea of the wave-cycle might repay careful

consideration as a basis for understanding the nature of defence production over extended periods of time. It is to that end that we analyse the role of state procurement in the formation of the British aircraft industry during the crucial 1914-45 era and note how the oscillations in that procurement policy had significant effects not only on the health of the industry as a whole but also on the regions hosting the production enterprises.

Demand for UK Military Aircraft

It is no exaggeration to claim that without government contracts the British aircraft industry scarcely would have survived during the 1914-45 period. Indeed, government orders as a result of World War 1 transformed the industry from a small-crafts shop basis, dominated by French designs and producing custom-built machines, into a major industry which by the time of the Armistice was, if not the biggest, then qualitatively the best anywhere. The figure of 34,147 aircraft ordered into production in the year following 1 November 1917 is illustrative of the gigantic wartime proportions of the aircraft industry. None the less, it is fair to say that state intervention in the industry was hesitant in the first place. On the one hand, the government fostered technical innovation through its large design staff at the Royal Aircraft Factory, Farnborough. That facility, along with pre-existing private firms (the list firms, selected according to whether they had previous ties with the War Office or Admiralty) was also responsible for aircraft production. On the other hand, the government was reluctant to establish further state-owned production plants. By mid-war, however, production could not keep up with demand and this was so despite the subcontracting of Farnborough designs to private manufacturers. Casting aside its reservations, the government responded by drawing upon the resources of organisations lacking in aviation expertise: among them furniture makers, piano makers, agricultural machinery makers and merchant shipbuilders. Table 1.5 lists some of the major enterprises inducted into wartime airframe manufacture. Co-opted firms either had to have an abundance of skilled workers who could turn their attention to aeronautical tasks without much ado, or had to have the abilities needed to handle large workforces which were used to following simplified engineering blueprints. At a time when wood and fabric materials dominated airframe assembly, furniture makers conformed to the former category while shipbuilders fell into the latter. Combining both attributes were members of the automotive industry. The two Birmingham firms of Austin and Wolseley concentrated on Farnborough designs, as did the Coventry firms of Daimler and Siddeley-Deasy. Other members of the West Midlands vehicles complex devoted their attention to producing the designs of private aircraft companies (i.e. Humber at Coventry and Sunbeam in Wolverhampton) while Standard Motors (of Coventry) built according to blueprints supplied by Farnborough and two private companies. A similar diversity in subcontracting was reflected in the output of co-opted vehicles firms in Lancashire and London.

In many respects government intervention in the aircraft industry echoed the philosophy of 'gradualism' which derived from the Ministry of Munitions. There, acceptance of total state control of production was resisted for some time by a reluctance to eschew laissez-faire economic ideology.[42] Various means of contracting directly to private arms firms or encouraging co-operative schemes among private subcontractors were given precedence over the erection of state-controlled munitions plants. The latter came into their own only when munitions needs overwhelmed private supply capacity and even then the new state facilities were managed by personnel loaned by the private sector. In a similar manner, aircraft supply became so imperative as to force the government to create several temporary production plants (it did the same with shipping supply) in the last year of the war (the Croydon, Heaton Chapel and Aintree plants actually coming on stream) and these National Aircraft Factories, likewise, were put under the auspices of private management: Aintree, for instance, was managed by the Cunard Steamship Company. The lacklustre commitment to a direct state role in aircraft manufacture was well illustrated by the decision taken in 1917 to confine Farnborough to basic research and to hand over its airframe and aeroengine designs to private firms.

In consequence, World War 1 set the pattern whereby the government would specify the type and quantity of aircraft required and leave the design and production of them to the private sector (admittedly, aided by technology furnished by the state). In this respect, the private aircraft firms were in a favourable position at war's end. However, the removal of competition following from the state's withdrawal from production was hardly sufficient to compensate for the drastic all-round termination in

Table 1.5 : World War I Airframe Entries

Firm	Location	Entry
Alliance Aeroplane	Hammersmith	engineering diversification
Austin Motors	Birmingham	motor vehicles diversification
Avro	Hamble	multi-plant extension
Armstrong Whitworth	Selby	horizontal move into airships
BAT	Willesden	1917 formation
Boulton & Paul	Norwich	diversifying building firm
Central Aircraft	Kilburn	wartime subcontractor
Fairey Aviation	Hayes and Hamble	1916 formation
Gosport Aircraft	Gosport	yacht builder's diversification
Hewlett & Blondeau	Leagrave	engineering diversification
London Aircraft	Lower Clapton	wartime subcontractor
Martinsyde	Woking	war-induced
Mann, Egerton & Co	Norwich	motor vehicles diversification
Nieuport & General Aircraft	Cricklewood	founded 1916 as branch of French firm
Parnall & Sons	Bristol	woodworking diversification
Phoenix Dynamo	Bradford	electrical diversification
Frederick Sage & Co	Peterborough	furniture diversification
S.E. Saunders	Cowes	boat builder's diversification
Standard Motor Co	Coventry	motor vehicles diversification
W.G. Tarrant	Byfleet	workworking diversification
Westland Aircraft	Yeovil	engineering diversification

Source: <u>Jane's All the World's Aircraft 1919</u> (Arco Publishing Co., New York, 1969).

military contracts. Many firms were unable to adjust
to peacetime conditions and either went bankrupt or
reverted to non-aviation products. For example, of
the firms highlighted in Table 1.5 as war entries,
the vast majority had withdrawn from the airframe
manufacturing field by the early-1920s while a small
number persevered for a few years longer (e.g.
Beardmore and Parnall). A few survived (and some
even prospered) from the judicious allocation of the
limited military contracts then issuing from the Air
Ministry. Only one or two firms were able to develop
viable civil aircraft. Turnaround in the industry's

fortunes came with the rearmament drives of 1934-7 initiated as a result of the failure of the Disarmament Conferences and the growth of aggressive forces in Italy and Germany. These measures meant not only expanded orders for existing firms, but the construction of new shadow factories of two kinds: those 'agency' type funded by the government and managed by the private airframe and aeroengine companies, and those provided by the government and managed by the main car firms (e.g. Austin, Ford, Rootes and the Nuffield Organisation). They were to be brought on stream at an increasing pace as and when the international scene deteriorated. Those of them not built beside existing aircraft industry sites were located according to three criteria: accessibility to pools of labour in established centres of engineering (particularly adjoining plants of the automotive industry), dispersal for strategic reasons, or placed in locations of high unemployment where the labour force would need adjusting to new skills.[43] Numerous plants in the West Midlands, London, Manchester and West Yorkshire attested to the first category, the location of the Shorts flying-boat factory at Windermere was an example of the second, whereas the decision to establish a bomber-assembly plant at Chester was an instance of the third. Sometimes, as in the case of Belfast, all three criteria seem to have been at work in influencing the government on where to invest in aircraft plants.[44]

With the onset of World War 2 special heavy-bomber production groups were established at government instigation to utilise the combined floorspace and organisational competence of automotive and electrical engineering firms and, ultimately, they accounted for 45 per cent of heavy bomber output. Thus, to supplement the shadow scheme, firms such as English Electric and Metro-Vickers were tasked with the production of bombers at Preston and Manchester respectively (Table 1.6). All told, the momentum of industry expansion continued until 1945. The ten years of unprecedented growth can be illustrated by the performance of two typical aircraft firms: Armstrong Whitworth Aircraft of Coventry and the Bristol Aeroplane Company. As a result of orders for bombers in 1935, the former increased its payroll from 760 in 1934 to 2,345 in 1936 and to about 5,000 in 1937. After the beginning of hostilities, it ran shadow factories at Leicester, Nuneaton, Northampton, Sywell and Bitteswell as well as its two Coventry plants and concluded the war with a workforce of over 12,500.[45] Bristol Aeroplane, meanwhile,

Table 1.6 : 'Shadow' Aircraft Production

Shadow firms[1]	Lead firm (designer)	Aircraft type
Metro-Vickers[2]	Avro	Lancaster
Austin[2]		(heavy-bomber)
Armstrong Whitworth[3]		
Vickers[3]		
Armstrong-Whitworth[3]	Avro	Lincoln
Metro-Vickers[2]		(heavy-bomber)
English Electric[2]	Handley Page	Halifax
London Aircraft[2]		(heavy-bomber)
Rootes[2]		
Fairey[3]		
Austin[2]	Short	Stirling
Short & Harland[3]		(heavy-bomber)
English Electric[2]	Handley Page	Hampden (medium-bomber)
Rootes[2]	Bristol	Blenheim (light-bomber)
Standard Motors[2]	D.H.	Mosquito
Airspeed[3]		(light-bomber)
Percival[3]		
Westland[3]	Vickers	Spitfire (fighter)
Austin[2]	Hawker	Hurricane
Gloster[3]		(fighter)
Rootes[2]	Bristol	Beaufighter
Fairey[3]		(fighter)

Notes: 1. Schemes involving non-aircraft firms only
2. Non-aircraft firm
3. Aircraft firm

Source: W. Hornby, Factories and Plant (HMSO, London, 1958), p. 244.

increased its payroll from 4,200 in June 1935 to
8,233 by December of the same year as a consequence
of retooling for the armament programme. In addition to its main plant at Bristol, the firm went on
to operate two airframe shadow factories at Weston-
super-Mare and Accrington, several aeroengine shadow
factories in the West Midlands, and employed 52,000
workers by 1942.[46] Even a formerly small subsidiary
of Vickers-Armstrongs, the Supermarine Company, was

employing more than 26,000 people by 1944 at South-
ampton, Swindon and Castle Bromwich. The climax to
expansion came in 1944 when over 1.5 million people
were employed by the aircraft industry as opposed to
35,000 in 1935.

Regional Shifts in Aircraft Production

As is evident from Table 1.7, the two major wars
stimulated the aircraft industry to output levels of
massive proportions. The fact that peacetime output
was but a fraction of wartime procurement is amply
demonstrated in the table, although even in the
Interwar era output fluctuated widely. The first
phase of that era, 1920-8, covers the effective
lifespan of equipment produced during World War 1
and also coincides with the time when official circles held the view that further large-scale war was
unthinkable. It is scarcely surprising, then, that
military procurement was minimal. The 1929-34 phase
monitors the time when wartime stocks were wearing
out and had to be replaced. It also occupies the
period when the national economy plummeted into
depression. Budget-paring in association with few
perceived threats on the international horizon were
not conducive to expenditures in military aviation
(although the substitution of 'air policing' for
expensive land force deployments in the Middle East
certainly provided the grounds for a larger air
force than would have been justifiable otherwise).
Conversely, the 1935-9 phase has something of an
analogy to the Cold War of later decades in which
rearmament was given utmost priority. When consid-
ered separately, therefore, the various phases of
the Interwar era had markedly different repercussions for the aircraft industry and cannot be
treated as a homogeneous contrast to the wartime
phases.
    Disaggregation of absolute numbers of aircraft
built for each phase gives some idea of the regional
distribution of the industry. As Table 1.7 indicates, the principal industrial centres have figured
prominently in this output, as indeed have smaller
centres located in Avon (focused on Bristol) and
Surrey. However, the table also serves to underscore
fluctuations by region as well as through time. So,
for instance, Greater London's share of aircraft
output swung from 27 per cent of the national total
in the earlier war to only seven per cent in the
later one, whereas the position of the West Midlands
remained virtually unchanged in both wars. We resort

Table 1.7 : UK Aircraft Production

	1914–19	1920–28	1929–34	1935–39	1940–45
Total no.[1]	66,140	3,851	3,643	21,818	119,777
Greater London	18,183	1,160	1,156	3,987	8,918
% of total	27	30	32	18	7
West Midlands	10,596	384	421	3,387	17,677
% of total	16	10	12	16	15
Greater Manchester	3,726	471	505	2,894	11,328
% of total	6	12	14	13	9
Avon	5,665	531	328	1,304	6,229
% of total	9	14	9	6	5
Surrey	4,208	257	476	824	4,409
% of total	6	7	13	4	4

Note: 1. Totals are derived from aircraft actually built from contracts placed during the specified years. There may have been some lag between the placing of contracts and the acceptance of the machine into service.

to a variant of shift-and-share analysis in order to systematise the fluctuations in aircraft output by regions over the 1914-45 period. This involves computing a total shift (TS) measure for each region:

$$TS = R_{t+1} - \left(\frac{N_{t+1}}{N_{t+1}} \cdot R_t \right)$$

where R is total aircraft production in a specified region during an initial phase (t) and subsequent phase (t+1), and N is aggregate national output corresponding to the same phases. The TS signifies the excess (a positive shift) or shortfall (a minus shift) in numbers of aircraft produced in the region in the recent time period relative to the region's share of national output established in the earlier time period. Accordingly, the magnitude of a +TS denotes the amount of extra output over and above

the region's tally if it had maintained a constant
competitive standing while, by way of contrast, .the
strength of a -TS is an indication of the level of
output foregone as a result of the eroding of the
region's competitive position. However, a qualifier
is in order: because each aircraft type has its own
battery of technological inputs, it is difficult to
equate output of one factory with output of another
producing a different design. A production run of
100 machines of one design may entail the same man-
hours of work as a production run of 200 of another
design (and that applies even with an assumption of
constant X-efficiency throughout the factories). As
a rule of thumb, we hold that aircraft size is a
good surrogate measure of the amount of technology
embodied in the aircraft: in short, the larger the
machine the more demanding its construction. Aeroengine capacity is taken as the size measure. Thus the
R and N terms in the formula are weighted by a factor which accounts for aircraft size (a twin-engined
machine having twice the weight of a single-engined
aircraft and so on). The resultant weighted total
shifts for 35 UK regions are presented in Table
1.8.[47]
    The left-hand column monitors the effects of
the onset of peace (t+1=1920-8) on the regions which
had established themselves as military aircraft suppliers in World War 1 (time t). Clearly, the main
benefactors in relative terms from the initial postwar contracts were Gloucestershire, Somerset,
Greater Manchester, Surrey, Avon and West Yorkshire
in that order, whereas the main losers were the West
Midlands, Lincolnshire, Norfolk, Strathclyde and
Tyne & Wear. Contrasting the t+2 phase (1929-34)
with the 1920s underscores the emergence of Hertfordshire and Surrey as leading producing areas
while, simultaneously, it points to the undermining
of West Yorkshire, Gloucestershire and Avon in the
competitive stakes. From 1935 rearmament was the
order of the day and the third column of the table
denotes its implications (t+3=1935-9). Now the magnitudes of the shifts are considerably enhanced and
they pinpoint widespread expansion in established
producing districts such as Hampshire, Berkshire,
Gloucestershire and the West Midlands as well as
newly-registered gains in places such as Northern
Ireland, Merseyside, Greater Manchester and Buckinghamshire. What is equally apparent, however, is the
massive relative declines in output experienced by
Greater London, Surrey and Somerset. Indeed, compar-
ing t+3 with t+4 (World War 2) further endorses the
relative decline of Greater London. Instead of

31

Table 1.8 : Weighted Total Shifts

Region	t→t+1	t+1→t+2	t+2→t+3	t+3→t+4	t→t+4
1. Central Scotland	- 9				- 483
2. Strathclyde	-148			1053	- 6897
3. N. Ireland	- 55		792	- 1910	1088
4. Cumbria				133	137
5. Tyne & Wear	-129				- 6946
6. Humber		47	531	- 5480	1193
7. W. Yorks	186	-237		11243	7573
8. Lancs .			142	16071	17179
9. Mersey	- 50		1192	- 512	5675
10. Greater Manchester	235	67	950	- 4358	18894
11. Cheshire			192	10098	11577
12. S. Yorks				210	214
13. Lincs	-235				-11624
14. Staffs				2393	2397
15. W. Midlands	-288	64	639	367	- 6754
16. Leics .	- 43			2368	86
17. Glos .	297	-176	891	- 7320	4994
18. Oxon				3450	3454
19. Wilts				1371	1375
20. Somerset	287	- 32	-1250	- 3902	249
21. Avon	191	-166	- 62	- 5847	- 5803
22. Cambs .	- 62				- 3320
23. Beds .	- 92		282	1087	- 1691
24. Norfolk	-177	5	- 339		-11779
25. Suffolk	- 39				- 2093
26. Hants	6	- 45	1953	- 8516	3611
27. Dorset				6283	6287
28. Herts		220	- 10	6792	19342
29. Greater London	46	88	-4487	-20732	-47165
30. Surrey	223	141	-3491	- 306	- 6400
31. Bucks			216	2802	4459
32. Berks			1323	- 5554	4306
33. E. Sussex	- 15				- 811
34. W. Sussex	- 13				- 721
35. Kent	- 58	7	469	- 1398	- 632

expansion in the major southeastern production
areas, the weight of war output swings to northern
regions such as Lancashire, West Yorkshire and Cheshire along with parts of the Midlands. This apparent leaning towards dispersal of plants perhaps for
strategic reasons is partly offset by the gains displayed by Hertfordshire, Buckinghamshire and Dorset:
regions certainly well within the comfortable reach

of German bombers.

The final column in Table 1.8 is a direct comparison of the regional distributions of aircraft procurement undertaken during the two World Wars. Four distinct trends can be elicited. The first is a relative decline in the importance of the Scottish and northeastern English producing areas in the Second as compared to the First World War. Next, the increasing importance of the northwestern, Yorkshire and south Midlands centres in the later war becomes evident. Thirdly, the collapse of eastern regions as major producing areas by the latter episode is especially noteworthy and, finally, dominating the scene is a massive relative fall in the competitive position of Greater London. In part, the shift in productive capacity is a reflection of the absolute decline of some regional economies, Strathclyde and Tyne & Wear for instance; but such an explanation will not suffice for either Greater London or Lincolnshire. Factors other than structural obsolescence have been at work in these cases. Among them was the relocation of aircraft factories from London to Hertfordshire and Buckinghamshire in the Interwar period so as to overcome restrictions on space for flying facilities, added to which was the deliberate attempt of government to site shadow factories away from the Metropolis. Also, the increasing complexity of airframe manufacture in which metal fabrication was substituted for wooden rigs endorsed the comparative advantage of major engineering agglomerations and their firms at the expense of small-crafts enterprises in Lincoln and East Anglia.

SUMMARY

Without doubt, therefore, fluctuations in demand can only be fully understood when, as here, local (regional) implications of production patterns are disentangled. This proviso applies equally to acquiring an effective appreciation of the supply factors underpinning defence industry. In both cases, moreover, such 'local' understanding awaits review of the behaviour of enterprises—the organisations which actually run the factories. Both of these concerns will be repeatedly dwelt on elsewhere in this book. Yet, the fact remains that comprehension of defence demand clearly resides in an awareness of the politics of international security and the wave-cycle notion is as good a basis as any for connecting that political dimension to the vicissitudes of defence production. The particularities of defence demand, as a theme, serves as the point of departure for the next chapter.

NOTES AND REFERENCES

1. For example, official DoD figures do not include defence-related activities of NASA or the Department of Energy. Consequently, US military expenditures may exceed the figure given in official statistics by up to 200 per cent. See J. Cypher, 'The basic economics of rearming America', Monthly Review, 23 (June 1981), pp.11-27.
2. Figures are derived from various SIPRI Yearbooks.
3. See introductory comments in J. E. Katz (ed.), The implications of Third World military industrialization, (Lexington Books, Lexington, Mass., 1986). Also germane here is the review of defence production in M. Brzoska and T. Ohlson (eds.), Arms production in the Third World, (Taylor & Francis, London, 1986).
4. Figures obtained from the US Office of Management and Budget, Federal outlays for major physical capital investment 1983, as manipulated by D. A. Nichols.
5. As extracted from the Statement of the defence estimates II, (Cmnd 9763-II, HMSO, London, 1986), p.17. Figures are expressed in outturn prices.
6. Cited in Aviation Week & Space Technology, 9 March 1987, p.227. For an expanded discussion of the military impacts of emerging technologies see F. Barnaby and M. ter Borg (eds.), Emerging technologies and military doctrine, (Macmillan, London, 1986).
7. Refer to Aviation Week & Space Technology, 20 October 1986, p.131.
8. The exceptions are Israel and the USSR.
9. For amplification of this dilemma in the UK instance, turn to D. Smith, The defence of the realm, (Croom Helm, London, 1980), p.120 and P. Pugh, The cost of seapower: the influence of money on naval affairs from 1815 to the present day, (Conway Maritime Press, London, 1986), especially Chapter 9.
10. The issue was publicised by N. R. Augustine. See his 'One plane, one tank, one ship: trend for the future?', Defense Management Journal, (April 1975), pp.34-40.
11. Refer to Aviation Week & Space Technology, 3 March 1986, p.23.

12. The discussion in this section is inspired by the work of Mary Kaldor.

13. See, in particular, K. Hartley, NATO arms co-operation, (George Allen & Unwin, London, 1983).

14. As discussed in K. Hartley, 'Public procurement and competitiveness: a community market for military hardware and technology', Journal of Common Market Studies, vol. 25 (1987), pp.237-47.

15. At the same time cost-plus contracts have declined from 22 to nine per cent. Note, Statement of the defence estimates II, p.14.

16. As recounted in US General Accounting Office, DoD's industrial modernization incentives program: an evolving program needing policy and management improvement, GAO/NSIAD-85-131, September 1985.

17. Properly styled the Report of the committee on inquiry into the aircraft industry, (Cmnd 2853, HMSO, London, 1964-5). Quotes from p.26 and p.92.

18. See commentary in the International Herald Tribune, 29 August 1986, pp.1, 15.

19. Cited in Interavia, April 1983, p.333.

20. Reported in Aviation Week & Space Technology, 25 February 1985, p.22.

21. Material presented here is derived from information supplied by Boeing International Corporation.

22. D. E. Weatherbee, 'Indonesia: its defense-industrial complex' in J. E. Katz (ed.), (1986) The implications of Third World military industrialization, pp.165-85.

23. GD is obligated to market the aircraft up to a value of $168 million. Indonesia is to receive offsets worth 35 per cent of the F-16 flyaway costs in the form of component subcontracts. See Financial Times, 23 March 1987, p.25 and Aviation Week & Space Technology, 22 June 1987, p.37.

24. D. Todd and J. Simpson, The world aircraft industry, (Croom Helm, London, 1986), Chapter 7.

25. G. Kennedy, Defense economics, (Duckworth, London, 1983), p.151.

26. See Shipbuilding inquiry committee 1965-66 report, (Cmnd 2937, HMSO, London, 1966), p.32.

27. N. S. Ross, 'Employment in shipbuilding and ship-repairing in Great Britain', Journal of the Royal Statistical Society, Series A, 115 (1952), pp.524-33.

28. British figures are taken from Table 6.1 of M. Kaldor, 'Technical change in the defence industry' in K. Pavitt (ed.), Technical innovation and British economic performance, (Macmillan, London, 1980), pp.100-121. US figures are derived from R.

W. DeGrasse, 'Military spending and jobs', Challenge, July-August 1983, p.10.
29. G. Adams, The politics of defense contracting: the iron triangle, (Transactions Books, New
Brunswick, 1982).
30. R. W. DeGrasse, Military expansion economic
decline: the impact of military spending on US economic performance, (M. E. Sharpe, Armonk, NY, 1983),
p.10.
31. H. F. Scott and W. F. Scott, The armed
forces of the USSR, (Westview, Boulder, Colo.,
1981), p.295.
32. D. L. Shambaugh, 'China's defense industries: indigenous and foreign procurement' in P. H.
B. Gordon (ed.), The Chinese defense establishment:
continuity and change in the 1980s, (Westview,
Boulder, Colo., 1983), pp.43-86.
33. M. Edmonds (ed.), International arms procurement: new directions, (Pergamon, New York,
1981), p.10.
34. Kaldor, 'Technical change in the defence
industry', p.102.
35. G. C. Allen, The industrial development of
Birmingham and the Black Country 1860-1927, (George
Allen & Unwin, London, 1929), pp.375-8.
36. U. Kobayashi, Military industries of Japan,
(Oxford University Press, New York, 1922), pp.184-5.
37. F. Abolfathi, 'Threat, public opinion, and
military spending in the United States, 1930-1990'
in P. McGowan and C. W. Kegley (eds.), Threats,
weapons, and foreign policy, (Sage, Beverly Hills,
1980), pp.83-133.
38. The information was gleaned from a variety
of sources, especially the series of works on warship evolution produced by the Conway Maritime
Press.
39. Strictly speaking, the shipbuilding interests of Vickers were nationalised in 1977 (and privatised again in 1986 with a management buyout). In
the period up to 1897 the Barrow shipyard was outside Vickers ownership. From 1927 it was complemented by the former Armstrong Whitworth naval shipyards on the Tyne (until these were relinquished to
Swan Hunter in the late-1960s). For its part, Mitsubishi includes the interests of Mitsubishi Zosensho
and Shin-Mitsubishi Jyuko at Nagasaki, Yokohama and
Kobe.
40. D. Todd, The world shipbuilding industry,
(Croom Helm, London, 1985), Chapter 5.
41. R. Higham, 'Quantity vs. quality: the
impact of changing demand on the British aircraft
industry', Business History Review, 42 (1968),

pp.443-68. For a generalisation of the concept see his _Air power: a concise history_, (St Martin's Press, New York, 1972).

42. As debated in R. J. Q. Adams, _Arms and the wizard_, (Texas A & M University Press, College Station, 1978).

43. Although, in practice, the defence ministries tended to disparage the regional policy connotations of strategic-industry location and were only interested in the depressed areas in as much as they fulfilled the requirement of being beyond German bombing range. See A. A. Lonie and H. J. Begg, 'Comment: further evidence of the quest for an effective regional policy 1934-1937', _Regional Studies_, 13 (1979), pp.495-500.

44. J. W. Blake, _Northern Ireland in the Second World War_, (HMSO, Belfast, 1956).

45. O. Tapper, _Armstrong Whitworth aircraft since 1913_, (Putnam, London, 1973).

46. C. H. Barnes, _Bristol aircraft since 1910_, (Putnam, London, 1964).

47. The data are computed from material contained in B. Robertson, _British military aircraft serials_, (Patrick Stephens, Cambridge, 1979).

Chapter Two

EMPIRICAL DIMENSIONS OF THE ARMS ECONOMY

A superficial review of some recent events in the
shipbuilding, aerospace and automotive industries is
sufficient to emphasise the extreme importance of
defence work to many enterprises and their workforces. From the shipbuilding viewpoint, the award
by the MoD in mid-1986 of three Type 23 frigate contracts worth £345 million for the Royal Navy was
reputed to guarantee 10,000 jobs over the next few
years. Each ship (and two were allocated to Yarrow
in Glasgow with the other going to Swan Hunter on
the Tyne) was said to require 1,200 man-years in
direct shipyard labour and twice as much again in
systems subcontracting and marine engineering. At
about the same time, the giant Newport News enterprise in Virginia announced record profits for the
sixth consecutive year ($285 million before tax on
sales of $1.8 billion, a far cry from the $14 million profits of 1979) and enjoyed an order book for
US Navy vessels worth in excess of $5 billion. The
company was building three Nimitz-class nuclear-powered aircraft carriers and eight Los Angeles-class
nuclear attack submarines with another carrier and
two submarines under refit. In total, 30,000 people
were employed by this shipbuilder, while the Hampton
Roads area had a further 8,000 employed in naval
ship-repair at other companies.[1] A good example from
the aerospace arena is provided by the receipt, in
mid-1986, of a £75 million order for rocket motor
casings by Lucas Aerospace. Part of the UK contribution to the four-nation multiple-launch rocket
system (MLRS), the order should ensure the survival
of the 1,500-strong workforce at Lucas's Burnley,
Lancashire, plant well into the 1990s and offers the
prospect of creating an extra 100 jobs.[2] Finally,
the UK also provides a suitable venue for illustrat-
ing the automotive industry case. In the late-1970s,
the RO (now Vickers-owned) tank factory at Leeds was

dealt a catastrophic blow when the downfall of the
Shah led to the cancellation of Iranian MBT contracts and slashed the work backlog from eight years
to a mere three months. It was only timely orders
worth £400 million for 280 Challenger MBTs for the
British Army and some replacement Jordanian Khalid
tanks that sustained the plant for the next four
years, albeit with a reduced labour force of 1,600.[3]
These examples are representative of defence industry in general and, in relating the degree of dependence of employer and employee alike on defence contracts, they hint at the reliance of entire
economies on such work. Not only are communities of
the likes of Glasgow, Newcastle, Burnley and Leeds
sensitive to defence budgets in as much as significant numbers of their local jobs derive from them,
but whole regions have come to rely on defence
industry for their very well-being. Indeed, the
well-being of nations--at least in terms of income
and employment effects--depends in no small measure
on the role of defence budgets in maintaining the
industrial system. In light of such stark reality,
the object of this chapter is to trace the economic
impacts of defence-industrial activity on the spectrum of economies, beginning with the national level
and extending, via the regional level, to that of
the local community. As a specific objective, the
chapter demonstrates that defence-dependence is
scale variant in a spatial sense, that is, defence
industry becomes relatively more important in an
economy with the progression down the size hierarchy
from nation to region to town.

AGGREGATE ECONOMIC IMPLICATIONS

The growth-inducing effects of defence production
are immersed in controversy. On the one hand, asser-
tions are made to the effect that defence purchases
create a large number of jobs: the figure bruited
abroad by the DoD is that on average about 35,000
jobs are generated with each billion dollars worth
of spending.[4] By implication, therefore, defence
spending is not only ensuring the public good
through enhanced security, but it is providing a
goodly proportion of the public with their livelihoods to boot. On the other hand, the opportunity
costs (i.e. alternatives foregone) of defence production put the defence industry in a much less com-
plimentary light. For example, the total labour
requirements for a $1 billion guided missile pro-
gramme was put at 53,248 which compares unfavourably

with equivalently-costed programmes in housing construction (68,657), public utility construction
(65,859) and the manufacture of either public transport equipment (77,356) or railway equipment
(54,220). A specific defence production programme,
the original ill-fated B-1 bomber of the Carter era,
was estimated, at its peak, of being capable of gen-
erating 58,591 jobs. Yet, for the same outlay, no
fewer than 118,191 jobs could have been created in
the education sector, not to speak of 108,196 in
social security if the money had been so directed,
or 99,406 in administering welfare payments and
88,415 in the conservation and recreation sector.[5]
Even more damning is the claim that defence production imposes crowding-out effects on industry. Thus,
it is charged that investment in defence industry
reduces the pool of available capital to such an
extent that legitimate civil industrial investment
is undermined and, what is more, the ground is cut
away from civil industrial R & D since firms choose
to concentrate on publicly-subsidised defence-industrial R & D. These imputations have been refuted (at
least in part) by the counter-claims that investment
seeks the greatest opportunity in a free market and,
therefore, investors will forsake civil fields if
they appear unpromising or uncompetitive. In any
event, defence R & D is replete with spin-offs (or
so it is argued) which invigorate civil production.
As a final and unimpeachable line of argument in its
favour, defence production is 'often taken as an
important built-in stabiliser to a depressed economy', which is to say, it provides a useful correc-
tive to cyclical downturns in business activity or,
possibly, an antidote to secular decline in established industries.[6]

The Counter-Cyclical Function

Circumstantial evidence certainly exists for the
counter-cyclical role of defence production. Severe
cutbacks in merchant ship production as a result of
business cycle downturns have often been met by
stepped up orders for naval construction.[7] While on
the surface totally distinct from business cycle
fluctuations, the oscillations of the wave-cycle can
be amended within limits to accommodate shortfalls
of mercantile output. In practice, this means that
governments can steer replacement contracts of the
kind that typify the peacetime equilibrium phase of
the wave-cycle to shipbuilding enterprises in espe-
cial need of topping up orders. A major constraint

on this activity is the degree of specialisation
undertaken by the units of the shipbuilding industry: quite simply, they may enjoy neither the diversity of plant nor labour skills to allow them to
readily switch to defence production during trade
cycle downturns. Nevertheless, the unemployment consequent upon recession in merchant shipbuilding in
the later 1970s forced the UK Government to forego
its declared policy of concentrating naval orders on
specialist warship yards and, instead, it expressly
disbursed them to erstwhile merchant yards at Newcastle and on Merseyside. Similarly, the West German Government responded to the same crisis by
spreading the orders for six frigates to five different yards even though such dispersion incurred
additional costs of DM 114 million.[8] Rather than
assign all six ships to the 'lead' yard of Bremer
Vulkan, other shipbuilders (AG Weser, Blohm + Voss,
Thyssen-Nordseewerke and Howaldtswerke (now HDW),
Kiel) were allowed shares in the programme in order
to sustain their workforces. Across the Baltic, the
virtual eclipse of shipbuilding in Sweden has been
prevented by the intervention of government as dis-
penser of warship contracts. Due to follow Uddevallavarvet into closure as a consequence of vanishing
merchant ship demand, the Kockums yard at Malmö is
to be retained after merchant shipbuilding ceases
there in 1988 solely on the strength of submarine
orders. It will join Karlskronavarvet and its workload of corvettes and submarines as the only remaining significant elements in the Swedish shipbuilding
industry.[9] This Swedish situation is simply of a
piece with the longstanding American practice of
substituting naval work during periods of scarce
mercantile activity. The harsh climate for merchant
shipbuilding in the 1980s is reflected in the lean
order books of US shipyards. As of July 1986, the
total deadweight tonnage of merchant ships building
or on order in the USA equalled 755,760 (Table 2.1).
Yet, as much as 34 per cent of that total was destined to go to the US Navy in the form of naval auxiliaries (i.e. some 69,120 furnished by GD and
185,400 by Avondale Shipyards).[10]

Of course, the counter-cyclical function is not
confined to the shipbuilding industry. As Table 2.2
makes clear, the recession in civil aircraft markets
which occurred after 1979 and was made manifest
through sharply reduced deliveries of commercial
airliners and private general aviation aeroplanes
conspired to drastically reduce activity levels in
the US aircraft industry. This trying period witnessed a drop in the number of completed airliners

Table 2.1 : US Merchant Ships on Order, 30 June 1986

Company	Location	Vessel type	Tonnage (dwt)	Customer
American Ship Building	Lorain, Ohio	tanker	30,000	US shipowner
Avondale Shipyards	New Orleans	6 tankers	185,400	US Navy
Bay Shipbuilding	Sturgeon Bay, Wis	3 container liners	48,390	US shipowner
GD	Quincy, Mass	2 logistics vessels	69,120	US Navy
National Steel	San Diego	2 tankers	418,000	US shipowner
Tacoma Boat	Tacoma	waste disposal vessel	4,850	US shipowner

Source: Tonnage figures derived from Fairplay, 17 July 1986, p. 30.

Table 2.2 : US Aircraft Production, 1978-84

	No. of civil airliners delivered (1)	No. of General Aviation aircraft delivered (2)	No. of military aircraft delivered to US forces (3)	Ratio of (3)/(1+2)
1978	241	17,817	723	.04
1979	376	17,055	734	.04
1980	387	11,881	819	.07
1981	387	9,457	918	.09
1982	232	4,266	758	.17
1983	262	2,688	836	.28
1984	185	2,438	624	.24

Source: Computed from information in Aerospace Facts and Figures 85-86
(Aerospace Industries Association of America, Washington, DC, 1985).

from 376 in 1979 to 185 in 1984, added to which was the plummeting of light aircraft deliveries from over 17,000 in the earlier year to 2,438 in 1984. In partial recompense, however, domestic US military aircraft deliveries remained fairly constant, indeed, even increasing slightly in the early-1980s. Proportionately, military aircraft increased their share of total aircraft deliveries from the four per cent figure of 1979 to about 25 per cent by the mid-1980s. Notwithstanding the crude nature of these production figures, they unequivocally imply that the aircraft industry can make use of defence production to fall back on when civil markets fail. Even the USA's northern neighbour cannot remain indifferent to this process. The Canadian subsidiary of Bell-Helicopter-Textron has been forced to abandon plans to design and produce a new model helicopter (the Model 400 TwinRanger) owing to the evaporation of civil markets: instead, underused capacity will be diverted to components manufacture for US military helicopters.[11]

Industrial Implications

Controversy aside, there is no gainsaying the importance of defence industry to the national economies of some AICs. From the profit side alone, it has been said that aerospace and defence accounts for about 22 per cent of the profits made by the UK engineering sector in comparison with the ten per cent contributed by electrical engineering, the nine per cent contributed by engineering components, and the five per cent advanced by mechanical plant contracting. In so far as employment is concerned, defence industry again is of some import. Putting aside the fact that the industry tends to 'hoard' labour, that is, retain it even when activity levels would not strictly justify such behaviour, defence production adds appreciably to the employment base of engineering industries.[12] In the UK, for instance, it was estimated as supporting 240,000 direct jobs (plus an extra 80,000 occupied in meeting arms export demands) and 330,000 indirect jobs in 1981, a standing that had improved noticeably over 1978 (when the equivalent figures were 219,000, 69,000 and 325,000 respectively) but was still inferior to the situation applying twenty years before (in 1963, for example, about 362,000 direct jobs were succoured through defence production).[13] France is in a roughly comparable situation, engaging about 271,000 workers in direct arms manufacture (of which 75,000 worked on exports) in 1976.[14] For the USA, however, the numbers are far greater. Direct and indirect employment resulting from defence production was put at 2,862,000 for 1980. As shown in Table 2.3, defence production

Table 2.3 : US Defence-related Employment, 1980

Industry	Total employment (1)	Defence-related[1] (2)	(2) as % of (1)
Aircraft	577,800	202,900	29.8
Guided missiles and space vehicles	93,400	73,100	64.2
Radio and TV communications equipment	419,800	183,300	43.6
Ordnance	82,600	32,400	39.2
Shipbuilding and ship-repair	170,000	65,800	36.2
Engineering and scientific instruments	49,700	5,300	10.0

Note: 1. Does not include defence export work.

Source: H.G. Moseley, The Arms Race, (Lexington Books, 1985), p. 51.

directly employed more than 200,000 people in the aircraft industry (and that is exclusive of export work), although proportionately, it is more important to the workforces of the guided missiles, communications equipment, ordnance and shipbuilding industries. Yet, above all, defence production is important to the national economy by virtue of its linkage effects. In other words, defence production contracts allocated to one industry have stimulatory effects on other industries acting as suppliers (or customers, if the contract entails semi-finished materials) to it. To be sure, these linkage effects are responsible for the creating of indirect jobs that stem from defence production, but even more significantly, they may encourage the entry of new industries into the economy. It is precisely owing to the specific demands of defence industry--demands that generally call upon technologically-intensive inputs--that a complex of engineering industries may be created, consolidated and sustained.

The potential linkage effects of defence industry, or, indeed, any industry, can be gauged by means of input-output analysis. At its simplest, this analytical tool describes the particulars of the inputs required to produce a unit of output throughout the processing sector.[15] The resultant

45

interindustry matrix is able to provide measures of
the relative weighting of every industry courtesy of
the assembly of input coefficients and their subsequent rearranging. Thus, associated with any industry is a vector of input coefficients or that array
of outputs from other industries that is incorporated into a unit of output of the industry in question. Each individual input coefficient is a constant value representing the proportion of industry
"i's" output absorbed as input into industry "j"
(the subject industry) relative to the total output
of "j".[16] A perusal of input coefficients fulfils
the purpose of identifying the principal supplier
industries serving a defence industry. As the current largest defence producer, the aerospace industry is an obvious candidate for input-output analysis. Utilising US data, it is possible to compile
vectors of input coefficients for the five constituent segments of the aerospace industry: aircraft,
aeroengines, propellers, aircraft equipment and complete guided missiles.[17]

The upshot is Table 2.4 and from it can be
deduced the fact that, regardless of segment, the
biggest single contributor to a unit value of aerospace output is the value added by the industry
itself: a telling indication of the technological
embodiment of aerospace production. Of bought-in
value, however, the aircraft segment is especially
beholden to the aircraft equipment segment and, secondly, to electronics suppliers (as subsumed under
the guise of the radio and TV equipment industry).
Much less important are suppliers in the nonferrous
metals, machinery, scientific instruments, wholesale
trade and business services sectors. The main inputs
used in aeroengines derive from other establishments
within the same segment, although the products of
machine shops act as an important supplementary.
Similarly, the propellers segment relies upon
intraindustry sources for much of its input needs,
but it is almost equally reliant upon supplies issuing from the iron and steel forgings industry. For
its part, aircraft equipment is chiefly reliant on
other branches of the aerospace industry, particularly the aircraft segment. Rather uncharacteristi-
cally for aerospace, the guided missiles segment has
little in the way of intraindustry purchases even
though it draws heavily on inputs from the aircraft
and aircraft equipment segments.

Ties between industries can be formalised by
application of the concepts of backward and forward
linkages. The former is the more direct of the two
and occurs within the vertical processing chain

Table 2.4 : US Aerospace Input Coefficients

Supplier Industries	Aircraft & Parts	Aircraft Engines & Parts	Propellers & Parts	Aircraft Equipment	Complete Guided Missiles
Complete Guided Missiles				0.0280	
Blast Furnaces & Basic Steel Products		0.0154	0.0113		
Iron & Steel Forgings		0.0176	0.0608		
Primary Nonferrous Metals		0.0184			
Aluminium Rolling & Drawing				0.0133	
Nonferrous Rolling & Drawing, n.e.c.	0.0125				
Nonferrous Wire Drawing & Insulating			0.0163		
Nonferrous Forgings		0.0215			
Screw Machine Products		0.0103			
Metal Stampings			0.0274		
Pipes, Valves & Fittings			0.0186	0.0129	
Fabricated Metal Products			0.0188		
Special Dies & Tools	0.0137	0.0291		0.0111	
Power Transmission Equipment			0.0141		
Machine Shop Products	0.0104	0.0946	0.0440	0.0200	
Lighting Fixtures					0.0109
Radio & TV Equipment	0.0760	0.0135		0.0193	0.0556
Aircraft				0.1850	0.2206
Aircraft Engines & Parts	0.0392	0.1457	0.0670	0.0399	0.0109
Aircraft Equipment	0.1506	0.0158	0.0397	0.0984	0.1856
Engineering & Scientific Instruments	0.0173				
Mechanical Measuring Devices	0.0135				
Wholesale Trade	0.0165	0.0166	0.0165		
Miscellaneous Business Services	0.0155	0.0195	0.0132	0.0138	0.0136
Business Travel, Entertainment & Gifts	0.0107	0.0124	0.0161		
Value-Added	0.4758	0.4023	0.4474	0.3652	0.4158

Source: Input-Output Structure of the US Economy 1963 (Washington, DC, 1969)

where finished goods are transformed from unfinished materials. It arises out of the enhanced demand for the output of industry "j" triggering the purchase by that industry of the output of industry "i", a supplier of inputs to "j". Therefore, expansion in industry "i" is directly attributable to the backward linkage effect of industry "j". Alternatively, forward linkage occurs in the latter stages of processing. In this instance, growth in output of industry "j" may work towards a reduction in its unit costs as a consequence of scale or learning-curve economies, and the benefits of such a reduction are passed on to industry "p" which uses "j's" output as intermediate inputs. Clearly, industries

47

at the finished end of the manufacturing process--
such as aerospace--are likely to have extensive
backward linkages to their suppliers of semifinished
materials but few forward linkages (of the materials
kind) owing to their preoccupation with completing
products mostly destined for final demand users.
Mindful of this qualification, attention is focused
on backward linkages and they are computed for the
principal US industries using Rasmussen's methodology.[18] That entails the concoction of, first, the
'total requirements' or the sum of the direct and
indirect requirements provided by all industries to
the unit output of a given industry ("Z.j"), secondly, the 'average backward link' or the total
requirements weighted by the number of interindustry
sectors ("1/m Z.j"), and, finally, the 'power of
dispersion' ("U.j").[19] The last is simply the average backward link for a given industry divided by
the average industrial requirement pertaining to all
industries. An industry's total requirements, then,
is a column multiplier which denotes the total input
requirements needed to accommodate an increase of
one unit of its final demand.

As elicited from Table 2.5, a motley collection
of industries appear prominent on the grounds of
total requirements; namely, meat products, poultry
and eggs, broadwoven fabric mills, motor vehicles
and, significantly, the defence-industrial sector of
tanks and components. The aerospace segments with
total requirements in the 1.7-1.9 bracket sit
squarely in the modal range of manufacturing industries. They fail to achieve the backward linkage
stature of tank manufacture but have about the same
stimulatory impact as other defence industries in
the ordnance and electronics (sighting and fire control equipment) fields, and distinctly greater
impacts than the shipbuilding industry. The average
backward link measurement, however, discounts undue
dependence on one or very few suppliers and is thus
a more meaningful indicator of the importance of
interindustry connectivity to a particular industry
than is "Z.j" alone. Again, the aerospace segments
appear in the middling ranks of industries, performing better than cement manufacture and shipbuilding
(each recording 0.0045) but not so well as, say,
motor vehicles (0.0064) or aluminium rolling
(0.0063). Furthermore, the aerospace industry has
average backward links that score no better than the
ordnance sectors and are unequivocally inferior to
that applying to tank manufacture. The final column
in Table 2.5 displays the "U.j" or power of dispersion values. A comparison using the power of

Table 2.5 : Backward Linkages for US Industries

Industry	Total Requirements (Z.j)	Average Backward Link (1/m.Z.j)	Power of Dispersion (U.j)
Poultry & Eggs	2.830	0.0078	1.435
Cotton	1.845	0.0051	0.936
Forestry & Fish Products	1.853	0.0051	0.939
Iron Ore Mining	1.780	0.0049	0.903
Coal Mining	1.531	0.0042	0.776
Crude Petroleum & Natural Gas	1.493	0.0041	0.756
New Construction, Residential Buildings	1.801	0.0050	0.914
Complete Guided Missiles	1.930	0.0053	0.978
Ammunition	1.940	0.0054	0.983
Tanks & Components	2.533	0.0070	1.284
Sighting & Fire Control Equipment	1.958	0.0054	0.993
Small Arms	1.878	0.0052	0.952
Small Arms Ammunition	1.831	0.0051	0.928
Other Ordnance	1.764	0.0049	0.894
Meat Products	3.198	0.0088	1.622
Broadwoven Fabric Mills	2.507	0.0069	1.272
Hosiery	2.091	0.0058	1.061
Sawmills & Planing Mills	2.065	0.0057	1.046
Wood Household Furniture	1.964	0.0054	0.996
Pulp Mills	2.122	0.0059	1.075
Industrial Chemicals	1.828	0.0051	0.927
Plastic Materials & Resins	2.125	0.0059	1.077
Drugs	1.802	0.0050	0.914
Petroleum Refining	2.133	0.0059	1.081
Cement	1.610	0.0045	0.814
Blast Furnaces & Basic Steel Products	1.872	0.0052	0.949
Aluminium Rolling & Drawing	2.283	0.0063	1.158
Fabricated Structural Steel	1.994	0.0055	1.011
Screw Machine Products	1.723	0.0048	0.873
Steam Engines & Turbines	1.881	0.0052	0.954
Machine Tools, Metal Cutting Types	1.733	0.0048	0.879
Computing Machines	1.640	0.0045	0.831
Motor Vehicles & Parts	2.313	0.0064	1.172
Aircraft & Parts	1.766	0.0049	0.895
Aircraft Engines & Parts	1.927	0.0053	0.976
Propellers & Parts	1.907	0.0053	0.967
Aircraft Equipment	1.964	0.0054	0.995
Shipbuilding & Repairing	1.627	0.0045	0.824
Air Transportation	1.562	0.0043	0.791
Wholesale Trade	1.348	0.0037	0.683

dispersion is valuable because it implies that industries registering "U.j's" in excess of unity have heavy dependence on the rest of the economy, that is to say, they have strong backward linkage effects. Only tank manufacture of the defence group exceeds this benchmark, joining some aspects of the food, clothing, paper, metal refining and automotive industries as key propagators of linkage effects. On the whole, defence industries are not distinguished by significant backward linkage contributions to the national economy. Moreover, their average forward links are likewise not outstanding. Computed as the mean of the sum of row (i.e. output) entries, Table 2.6 shows that average forward links applying to the aerospace industry emerge generally lower than the magnitudes associated with that industry's average backward links. In short, excepting the notable case of tank manufacture, defence production has only modest stimulatory materials linkage effects on manufacturing at large in contrast to the classic key industry of motor vehicles.

However, because defence industries are typically technologically-intensive ('high-tech'), their importance to national economies is not confined to materials linkages of the input-output kind. Put bluntly, they afford access to sophisticated technologies which have far-reaching strategic, prestige, innovation spin-off, counter-cyclical demand and income effects: a set of attributes which, as made evident in Chapter 1, panders more to political than economic considerations. Consequently, the 'political' significance of defence industry at the national level tends to overshadow the pure interindustry linkage effects. The same, though, cannot be said for the role of defence industries in regional economies within nations, a theme to which we now turn.

REGIONAL IMPACTS

In an advertising promotion released in 1986 the US state of Georgia made a pitch for aerospace firms partly on the basis of the state's 'strong voice in Washington' which 'makes it obvious that the defense industry is welcomed and supported here'.[20] If for no other purpose, such advertising endorses the importance of defence industry to the regional economies of the USA and hints at the lengths to which the regions will go in order to influence the polity in a way which ensures that defence contracts are disbursed in their favour.[21] Melman has argued, in no uncertain terms, that the self-interest of the regions in defence production has colluded with other sectional interests to maintain American defence budgets at high levels.[22]

Sustained operation of the military economy has produced a major concentration of income flows

Table 2.6 : Aerospace Linkage Traits

Industry	Total Requirements (Z.j)	Average Backward Link (1/m.Z.j)	Average Forward Link (1/m. Z.i)
Complete Guided Missiles	1.930	0.0053	0.0029
Aircraft	1.766	0.0049	0.0044
Aircraft Engines & Parts	1.927	0.0053	0.0050
Aircraft Propellers & Parts	1.907	0.0053	0.0029
Aircraft Equipment, n.e.c.	1.964	0.0054	0.0061

in certain industries, occupations and regions. Thus the newly expanding military industries received major allocations of capital; their employees received exceptionally high pay and rapid promotions; and the geographic regions in which all this occurred underwent unusual 'economic growth'.

Interestingly, the main regional beneficiaries of the 'military economy' have not remained constant. Examination of industrial structure has shown that certain states, historically bypassed by defence production, have received growth stimulus since 1950 as a direct result of extensive defence purchases. Thus, the six western states of Kansas, Washington, California, New Mexico, Arizona and Utah were major gainers in manufacturing employment during the 1950s, and by 1960 defence-related employment amounted to over 20 per cent of all manufacturing employment in them.[23] As a point of fact, the western part of the USA has increasingly benefited from defence production at the expense of other parts of the country and this garnering of a new industrial fabric has been responsible in no small measure for the growth of it and the so-called 'Sunbelt' at large.[24] Table 2.7 intimates that three 'Sunbelt' regions—the South Atlantic, South Central and Pacific—have steadily increased their share of prime contract awards from a World War 2 average of 28.3 per cent of the total by value, through a Korean War average of 31.9 and a 1961 figure of 45.9, to a 1981 tally of 54 per cent. In a corresponding manner, the old industrial regions in the 'Frostbelt' have witnessed decline: the Middle Atlantic has plunged from 23.6 per cent of World War 2 contracts to 13 per cent of those disseminated in 1981 while the East North Central has gone from a position of enjoying about one-third of the prime contracts to having to make do with just nine per cent of them.

Only New England of the traditional industrial regions has retained (and, indeed, improved) its share of such contracts. Yet, from a position of rough equality with the Pacific region immediately prior to World War 2, it had regressed by the early-1980s to a stature of only half the latter's. The reason for the historical shift in the localisation of defence production is not hard to find. As technological intensity has gained ground in weapons systems, so too has the onus shifted to aerospace products and away from the less technologically-endowed products of vehicles, ships, ordnance and other ammunition. In short, boosted procurement of aircraft, missiles and electronics has occurred through budgets detrimental to traditional weapons platforms. The growing group embraces items which the Pacific and other western regions are particularly well qualified to produce--and in which New England retains some comparative advantage--whereas the declining group takes in items that originate by and large from the 'Frostbelt'. One dramatic instance of the geographical consequences of the shifting composition of defence procurement is Colorado. Between 1952 and 1962, this state increased its share of defence production by a factor of five: a growth rate which stemmed almost entirely from the action undertaken during this period by Martin Marietta that resulted in the establishment of a large missile plant at Denver.[25]

Table 2.8 offers some insights into the defence dependency of states at about the time of peak Vietnam War contracts. A considerable number of states' workforces have ratios higher than the national average of 3.6 per cent of all workers relying on defence production for their livelihoods. No less than 10.3 per cent of the workforce in the District of Columbia are defence dependent, and it is followed in short order by Utah (9.9), Alaska (9.8), Hawaii (8.8), Virginia (8.4), Connecticut (7.5), Maryland (6.9), California (6.5) and New Hampshire (6.4). Of course, not all of this employment is engaged in defence production: only Connecticut and California of the states inordinately reliant on defence employment are largely involved in defence

Table 2.7 : Regions and US Prime Contracts

Region	Annual averages (%)			Prime contract value ($ Million)	
	1941-5	1951-3	1961	1981	%
New England	8.9	8.1	10.5	10 357.1	12
Middle Atlantic	23.6	25.1	19.9	11 157.7	13
East North Central	32.4	27.4	11.8	7 637.8	9
West North Central	5.6	6.8	5.8	7 091.7	8
South Atlantic	7.2	7.6	10.6	12 482.6	15
South Central	8.8	6.4	8.2	13 849.4	16
Mountain	1.2	0.7	5.7	3 057.9	4
Pacific	12.3	17.9	26.9	19 331.9	23
Alaska and Hawaii	-	-	0.6	914.1	<1

Source: Derived from A. Buckberg, 'Federal Government Expenditures', The Western Economic Journal, spring 1965, Table 2 and C.H. Anderton and W. Isard, 'The Geography of Arms Manufacture' in D. Pepper and A. Jenkins, (eds.) The Geography of Peace and War, (Basil Blackwell, Oxford, 1985), Table 5.3.

production. Indeed, the ranking of states is subtly rearranged when the relative size of defence production contracts is taken into account. Weighting of 1981 prime contract awards by the magnitude of state workforce gives the final column entries in Table 2.8. As well as the nation's capital, California and New England (Connecticut and Massachusetts), a clutch of states in the 'Sunbelt' (Louisiana, Mississippi and Virginia) are complemented by such outposts as Alaska, Hawaii, Maryland, Missouri, Texas and Washington. In order to bring home the importance of defence production to these states, it is only necessary to review the effects of defence cutbacks. For example, a 30 per cent decrease in the 1975 defence budget was estimated as having the effect of boosting national employment by around two per cent over the succeeding five years on the assumption that the withheld expenditures were reallocated to domestic programmes. However, while the Middle Atlantic states of Pennsylvania and New

Table 2.8 : Defence-Dependent Employment, US States

State	Defence-dependency ratio (June 1967)	Proportion of defence employment pertaining to Prime Contractors (%)	Contract awards per worker 1981 ($)
Alabama	4.2	30	509.16
Alaska	9.8	28	1819.91
Arizona	4.2	52	902.61
Arkansas	1.7	29	212.33
California	6.5	61	1417.44
Colorado	3.4	40	613.52
Connecticut	7.5	95	2828.36
Delaware	1.3	40	825.82
District of Columbia	10.3	13	2008.31
Florida	3.5	63	702.29
Georgia	4.8	42	513.94
Hawaii	8.8	9	1315.69
Idaho	0.4	33	95.48
Illinois	1.8	42	218.17
Indiana	3.1	65	656.34
Iowa	1.2	92	234.99
Kansas	3.3	76	827.51
Kentucky	1.9	9	245.02
Louisiana	1.7	39	1639.81
Maine	1.9	70	935.97
Maryland	6.9	42	1104.75
Massachusetts	4.3	70	1555.20
Michigan	1.4	62	413.23
Minnesota	2.2	88	554.58
Mississippi	3.1	73	1371.39
Missouri	4.5	68	1903.96
Montana	2.7	33	113.79
Nebraska	1.4	22	141.87
Nevada	1.6	10	169.29
New Hampshire	6.4	32	818.91
New Jersey	3.3	62	646.52
New Mexico	4.5	27	670.90
New York	2.1	66	813.44
North Carolina	2.0	60	294.93
North Dakota	1.3	48	346.08
Ohio	2.5	54	486.05
Oklahoma	4.4	17	455.60
Oregon	1.0	38	128.82
Pennsylvania	3.2	39	439.31
Rhode Island	5.3	34	493.64
South Carolina	3.0	20	340.14
South Dakota	0.7	38	169.92
Tennessee	2.8	75	246.95
Texas	4.3	49	1060.63
Utah	9.9	33	626.09
Vermont	2.1	95	644.63
Virginia	8.4	29	1389.16
Washington	4.3	51	1405.58
West Virginia	1.5	78	145.60
Wisconsin	1.4	80	255.87
Wyoming	0.9	14	256.12
USA	3.6	54	-

Source: First two columns abstracted or derived from R.F. Riefler and P.B. Downing, 'Regional Effect of Defense Effort on Employment', Monthly Labor Review, July 1968; third column taken from R.W. DeGrasse, Military Expansion Economic Decline, (M.E. Sharpe, Armonk, 1983).

Jersey, along with New York, Illinois, Indiana and
Wisconsin, would be net beneficiaries from arms
reduction, both California and the Mountain region
would most definitely lose jobs.[26]
Lest the perception remain that disparities in
the regional significance of defence production
appertain only to the superpowers, it is necessary
to be alerted to the fact that they apply equally to
modest powers. To this end, Table 2.9 assigns Canadian defence-generated GDP (a measure combining
direct effects and the various multipliers) to its
provincial sources. Surmounting all provinces by far
is Ontario, responsible on its own for virtually
one-half of Canada's defence production. Its only
real competitor is Quebec, but this province is good
for less than one-fifth of the national defence production. Nevertheless, the smaller provinces with
weak industrial economies are relatively more dependent on defence industry than are the two large
industrialised provinces. Thus, the defence industry
share of GDP in Nova Scotia is proportionately three
times greater than is the case in Ontario and more
than four times greater than the share of defence
industry in Canada's GNP. It does not take much fore-
thought to appreciate that economies such as Nova
Scotia's are extremely sensitive to fluctuations in
defence budgets. In fact, the vital role of defence
production in marginal regions on the one hand, combined with the overwhelming dominance of Ontario as
a location of defence industry on the other, has
received some recognition in Canada. To recount the
official line, the idea is that in order 'to assist
in the attainment of the Government's objective of
regional economic equality, further decentralisation
of defence procurement into all regions of Canada
will be encouraged whenever this can be done consistent with long-term economic efficiency'.[27] In other
words, the Canadian authorities have expressed a
wish to see development in the marginal regions fostered by means of augmented defence industry,
although as we shall see in the next chapter, this
laudable aim is quickly dumped if it interferes with
the bail-out imperative. Regional policy notwithstanding, decentralisation of defence industry has
received little encouragement in the UK. A half century of official regional aid, not to mention wartime plant locations for 'strategic' reasons, has
failed to alter the skewed distribution of prime
contract allocations in Britain. The South East,
South West and East Midlands regions receive more of
the contracts than their share of national population would justify: the marginal regions of the

Table 2.9 : Defence Production, Canada 1982-3

Province	Defence-generated GDP ($ 000)	Share of defence production (%)	Defence share of Provincial GDP (%)
Newfoundland	34 324	0.7	0.7
Prince Edward Island	29 047	0.6	2.9
Nova Scotia	475 615	10.0	5.7
New Brunswick	159 042	3.4	2.4
Quebec	826 986	17.5	1.0
Ontario	2 206 771	46.6	1.6
Manitoba	177 718	3.7	1.3
Saskatchewan	69 477	1.5	0.4
Alberta	362 240	7.7	0.7
British Columbia	373 391	7.9	0.8
Territories	15 604	0.3	1.3
CANADA	4 730 215	100.0	1.3

Source: J.M. Treddenick, 'Regional Impacts of Defence Spending' in B. MacDonald (ed.), Guns and Butter: Defence and the Canadian Economy, (CISS, Toronto, 1984), Table 5.

country, by way of contrast, have to make do with less than their 'fair' share.[28] To make matters worse, the UK marginal regions tend to specialise in products emanating from the relatively 'low-tech' end of defence industry (e.g. ships' hulls): a characteristic which does not augur well for future growth prospects.[29] In a manner reminiscent of the US case, it is the regions in Britain replete with aerospace and electronics manufacturers that are the winners from defence procurement. Unlike the American situation, however, they are found for the most part in the national 'core' around the capital.[30]

LOCAL DEPENDENCE

The chapter began with random examples of recent
defence contracts and pondered the employment consequences of them, not least on the communities that
played host to the defence plants. No better symbol
of the symbiosis between defence work and community
well-being can be found than the isolated Cumbrian
town of Barrow-in-Furness. At its peak in 1921, no
less than 42.8 per cent of its workforce of 25,799
(in a population of 74,244) were employed in the
metal trades of engineering and shipbuilding--trades
dominated by the firm of Vickers.[31] We have already
singled out Vickers as an archetypal naval defence
contractor and indicated that the firm's warship
output declined dramatically after World War 1.
Concurrently, its host community of Barrow suffered
a 10.8 per cent drop in population between 1921 and
1931 (a precipitate decline exceeded only by the
coal-mining Rhondda and iron-making Merthyr Tydfil
in South Wales) which was a fundamental reversal of
the town's experience of boom conditions during the
war (as reflected in a population increase of 16.4
per cent between 1911 and 1921). The immediate aftermath of that war with the evaporation of work at
Vickers had the dire outcome of driving male unemployment in the town to an unprecedented 44.1 per
cent in 1922.[32] Sixty years later Barrow remains a
'one-industry' town in which the heirs to Vickers--
VSEL--employ about 9,000 shipyard workers and 4,000
engineering workers at a site which continues as the
UK's largest warship yard.[33] Indeed, Barrow exempli-
fies the strengths and weaknesses of towns and cities overly reliant on defence industry: expansion
and prosperity during the phases of rearmamental
instability and wartime equilibrium standing in for
the former, while testimony for the latter is provided by severe contraction in the demobilisational
instability phase and marking time when peacetime
equilibrium is in vogue.
There is no shortage of communities exposed to
this vulnerability. Some of them are 'defence
enclaves', that is, settlements created by defence
industries to fulfil the sole function of servicing
defence establishments. A swathe of territory in Los
Angeles County is dotted with such communities;
namely, Hawthorne, El Segundo, Culver City, Burbank,
Canoga Park, Pomona and Palmdale. They have since
been joined by Anaheim and Fullerton in the more
'suburban' Orange County. Elsewhere in California,
San Diego and parts of Santa Clara County ('Silicon
Valley') also qualify as defence enclaves. Not to be

outdone, Fort Worth, Texas, has seen a string of
defence-dependent suburbs spring up at Arlington,
Richardson, Irving and Grande Prairie as the aerospace industry has taken root there. Florida, too,
has its defence enclaves centred on the Cape Canav-
eral-Kennedy Center-Patrick AFB space complex
(Titusville, Cocoa Beach and Melbourne) as well as
discrete locations at Orlando, St Petersburg, Tampa
and West Palm Beach which manage to coexist quite
contentedly with cheek-by-jowl tourist resorts. All
of these communities are well-stocked with defence
electronics and missiles plants.[34] The space complex
alone--stemming from a modest missile test centre--
was responsible for a climb in local employment from
1,000 in 1950 to 38,000 in 1965.[35] In fact, defence-
dependency is no respecter of community size in the
USA, encompassing Los Angeles at one extreme and a
number of non-metropolitan cities such as Worcester,
Massachusetts, and Lakeland, Florida, on the
other.[36] Los Angeles was estimated as being dependent upon defence and aerospace activity for 43.5
per cent of its employment while, up the coast, the
San Francisco area owed 15.9 per cent of its jobs to
the same sectors. Also in the Pacific region, Seattle had some 42.2 per cent of its employment related
to defence spending.[37] Less striking but locally
significant, none the less, is Bristol in the UK.
More than 21,000 people or 7.3 per cent of employment in this English metropolitan area were working
for the aerospace industry (as against a national
average of 0.9 per cent) and that industry, for its
part, was predominantly defence related.[38]

While hard figures are difficult to amass,
Table 2.10 does serve the purpose of identifying
other British cities and towns which would appear to
complement Bristol by falling into the high defence-
dependency category. The actual sensitivity of these
towns to defence spending is contingent, of course,
on the relative weighting of defence in the work-
loads of the firms present within them. Obviously,
the greater the number of defence contractors in a
community the greater the chance that some of them
will be able to find civil orders to mitigate the
reliance on defence work. In Bristol's case, for
example, the two main aerospace defence contrac-
tors--BAe and Rolls-Royce--vary in their ability to
blunt the dependence on defence work. BAe is able to
blend maintenance of US F-111 bombers with parts
manufacture for civil BAe 146 and Airbus airliners
at its Filton plant whereas Rolls-Royce reserves its
Bristol site mainly for work on the RB.199, Adour,
Pegasus and Viper engines that are geared to

Table 2.10 : Defence Plants in UK Communities

Community	Region	Number of sites	Number of defence firms
Chelmsford	South East	12	1
Coventry	West Midlands	11	6
Bristol	South West	8	6
Portsmouth	South East	8	2
Edinburgh	Scotland	6	2
Glasgow	Scotland	6	5
New Malden	South East	6	1
Preston	North West	6	2
Walford	South East	6	5
Birmingham	West Midlands	5	1
Bracknell	South East	5	3
Yeovil	South West	5	1

Source: P. Southwood, 'The UK Defence Industry', Peace Research Reports No. 8, Bradford University, September 1985, p. 39.

military programmes. Currently, the RB.199 powers the Tornado, the Adour is the powerplant for the BAe Hawk jet trainer, the Pegasus is the prime mover for the BAe/MD Harrier V/STOL fighter, while the Viper is an older engine of the turbojet type used in several foreign (particularly Italian and Yugoslav) aircraft. As a generalisation, it follows, therefore, that sensitivity to defence budgets is enhanced in those communities devoid of alternative production opportunities, a situation likely to crop up where a single defence contractor dominates the local economy.

SUMMARY

A superpower economy such as the USA supports a defence-industrial base which directly employs just under three per cent of the civilian workforce (and as much as 13.6 per cent if all multiplier effects

of DoD spending are taken into account). A typical
US state without an undue endowment of defence
industry--Iowa--has at least three per cent of its
workforce specifically employed in defence-related
industry.[39] Yet, one Iowa town, Burlington, has virtually its entire workforce dependent on the agency-
operated Iowa Army Ammunition Plant with its 4,000
jobs whereas another, Denison, exhibits an eleven
per cent dependence on defence contracts even though
those contracts are furnished by a meat packing
firm! In an undramatic fashion, these statistics
substantiate the notion that defence-dependence
tends to operate in an inverse relation with geographic size; which is to say, the smaller the place
the greater its sensitivity to fluctuations in
defence demand. While not startling in itself, such
an asseveration is useful in that it focuses attention on the parties most affected by defence industry; namely, the communities that have a symbiotic
relationship with the production units of defence
industry and which bear the brunt of shifting
defence programmes in terms of employment generation
or lay-offs. In other words, the economic implications of defence industry are played out most
starkly in the 'defence enclave' type of community
and the future of such communities cannot be
divorced from questions of viability associated with
the local defence plant. Such issues, in turn, are
inseparable from questions concerning the role of
individual defence plants within their proprietary
industrial organisations. It is to the world of
industrial organisations and their defence manifestations that we devote the succeeding chapters.

NOTES AND REFERENCES

1. Recorded in two successive issues of Lloyd's
List; namely, for 16 and 17 July 1986.
2. See The Engineer, 12 June 1986, p.84.
3. See The Engineer, 17 March 1983, p.8.
4. As mentioned in Aviation Week & Space Technology, 11 February 1985, p.19.
5. H. G. Moseley, The arms race: economic and
social consequences, (Lexington Books, Lexington,
Mass., 1985), pp.97-101.
6. M. Seagram, 'Does relatively high defence
spending necessarily degenerate an economy?', Journal of the RUSI for Defence Studies, 131 (1986),
pp.45-9; quote from p.48.
7. For details refer to D. Todd, The world
shipbuilding industry, (Croom Helm, London, 1985),

Chapter 7.
8. M. Brzoska, 'The Federal Republic of Germany' in N. Ball and M. Leitenberg (eds.), The structure of the defense industry, (St Martin's Press, New York, 1983), pp.111-39.
9. The large merchant building yards have become progressively phased out. Commencing in the 1970s with Lindholmens, the Swedish Government through its Swedyards holding company, went on to eliminate new construction facilities at Eriksberg, Cityvarvet, Finnboda, Oresundsvarvet and Uddevallavarvet. See Fairplay, 12 June 1986, pp.17-18.
10. Calculated from data in Fairplay, 17 July 1986, p.30. Note that the 254,520 dwt earmarked for the US Navy exceeds the entire merchant ship order book of France (213,000) and comes close to equalling those of Belgium (269,720), Finland (272,520) and the UK (284,800).
11. The reapportionment of work was necessary under Bell's commitments to the Canadian Government. In moving the JetRanger and LongRanger civil helicopter lines to Mirabel, near Montreal, the company's Fort Worth, Texas, plant had more space to work on the V-22 programme for the US military. These transfers were unable to make full use of Mirabel capacity, however, and the production of tail booms for military UH-1 helicopters had to be relocated to Canada from Amarillo, Texas. See Aviation Week & Space Technology, 21 July 1986, p.27.
12. K. Hartley and W. J. Corcoran, 'Short-run employment functions and defence contracts in the UK aircraft industry', Applied Economics, 7 (1975), pp.223-33. Note, the 'hoarding' tendency is much less prevalent in the USA.
13. As revealed in The Engineer, 14 July 1983, pp.29-50. By 1984, however, the MoD appeared to be supporting 250,000 direct jobs in defence industry. See The Engineer, 23/30 August 1984, p.36.
14. P. Y. Hammond, D. J. Louscher, M. D. Salomone and N. A. Graham, The reluctant supplier: US decision-making for arms sales, (Oelgeschlager, Gunn & Hain, Cambridge, Mass., 1983), p.223.
15. An excellent review is W. H. Miernyck, Elements of input-output economics, (Random House, New York, 1965).
16. To be specific, the fundamental relationship of input-output analysis is:

$$X_i - \sum_{j=1}^{n} x_{ij} = Y_i$$

61

which implies that total output of industry "i", exclusive of "i's" output absorbed by each industry "j", equals the final demand output. Individual input requirements per unit output are derived from the ratio:

$$a_{ij} = x_{ij} / X_j$$

which, on substitution into the first relationship, gives:

$$X_i - \sum_{j=1}^{n} a_{ij} X_j = Y_i$$

The resultant interindustry matrix can be used to infer input effects of each industry by column, and output effects of each industry by row.
    17. The data is taken from Input-output structure of the US economy 1963, (US Department of Commerce, Office of Business Economics, Washington, DC, 1969); still the only comprehensive source of aerospace industry technical transactions.
    18. P. N. Rasmussen, Studies in inter-sectoral relations, (Einar Harcks Forlag, Copenhagen, 1956). For an upgrading of the methodology, see B. ÓhUalla- cháin, 'Input-output linkages and foreign direct investment in Ireland', International Regional Science Review, 9 (1984), pp.185-200.
    19. In detail, the total requirements is a column multiplier derived from the inversion of a vector of input coefficients:

$$\sum_{i=1}^{m} Z_{ij} = z._j$$

where "Z" is the inverse matrix $(I-A)^{-1}$. The average row multiplier, that is,

$$1 / m\, Z._j \quad (j = 1, 2, 3 \ldots m)$$

can be deduced to complement the average column multiplier (i.e. the average backward link) and the two are combined and normalised by the overall average:

$$1/m^2 \sum_{j=1}^{m} \sum_{i=1}^{m} Z_{ij} = 1/m^2 \sum_{j=1}^{m} Z_{\cdot j} = 1/m^2 \sum_{i=1}^{m} Z_{\cdot i}$$

On conversion into a ratio, this relationship provides an index,

$$U_j = (1/m \, Z_{\cdot j}) / (1/m^2 \sum_{j=1}^{m} Z_{\cdot j})$$

that is, the power of dispersion.

20. As found in Aviation Week & Space Technology, 7 April 1986, p.40.

21. Although in a statistical sense the accusation of 'pork-barrelling' can be substantiated, it does not appear to be the main factor accounting for the distribution of defence contracts. See R. J. Johnston, 'Political influences on the allocation of federal money to local environments', Environment and Planning A, 10 (1978), pp.691-704 and 'Congressional committees and the inter-state distribution of military spending', Geoforum, 10 (1979), pp.151-62. Notwithstanding the national scene, the proviso remains, of course, that pork-barrel contracts may be significant for individual regions with powerful political representation.

22. S. Melman, The permanent war economy: American capitalism in decline, 2nd edn (Simon & Schuster, New York, 1985), p.132.

23. J. W. Dyckman, 'Some regional development issues in defense program shifts', Journal of Peace Research, 1 (1964), pp.191-203.

24. E. J. Malecki, 'Military spending and the US defense industry: regional patterns of military contracts and subcontracts', Environment and Planning C, 2 (1984), pp.31-44.

25. R. E. Bolton, Defense purchases and regional growth, (The Brookings Institution, Washington, DC, 1966), p.122.

26. R. H. Bezdek, 'The 1980 economic impact--regional and occupational--of compensated shifts in defense spending', Journal of Regional Science, 15 (1975), pp.183-97.

27. Abstracted from a background paper prepared for the Western Economic Opportunities Conference of 25 July 1973 on behalf of the Hon Jean-Pierre Goyer, then Minister of Supply and Services for Canada.

28. J. Short, 'Defence spending in UK regions', Regional Studies, 15 (1981), pp.101-110.
29. C. M. Law, 'The defence sector in British regional development', Geoforum, 14 (1983), pp.169-84.
30. Although as noted in the previous chapter, London has lost much of its original comparative advantage in the production of military airframes-- the weapons platform aspect of aerospace manufacture--to the Outer Metropolitan area surrounding it.
31. T. H. Bainbridge, 'Barrow in Furness: a population study', Economic Geography, 15 (1939), pp.379-83.
32. F. Barnes, Barrow and district, (James Milner, Barrow, 1951), p.116.
33. See The Engineer, 28 January 1982, p.29.
34. A. R. Markusen and R. Bloch, 'Defensive cities: military spending, high technology, and human settlements' in M. Castells (ed.), High technology, space, and society, (Sage, Beverly Hills, 1985), pp.106-120.
35. D. C. Weaver and J. R. Anderson, 'Some aspects of metropolitan development in the Cape Kennedy sphere of influence', Tijdschrift voor Econ en Soc Geografie, 60 (1969), pp.187-92.
36. A. Markusen, 'The military remapping of the United States', Built Environment, 11 (1985), pp.171-80.
37. C. M. Tiebout, 'The regional impact of defense expenditures: its measurement and problems of adjustment' in R. E. Bolton (ed.), Defense and disarmament: the economics of transition, (Prentice-Hall, Englewood Cliffs, 1966), pp.125-39.
38. M. Boddy and J. Lovering, 'High technology industry in the Bristol sub-region: the aerospace/defence nexus', Regional Studies, 20 (1986), pp.217-31.
39. A. R. Sunseri, 'The military-industrial complex in Iowa' in B. F. Cooling (ed.), War, business, and American society, (Kennikat, Port Washington, NY, 1977), pp.158-70.

Chapter Three

THE CONCEPTION OF THE MIE

Hitherto, we have focused on defence industry in
aggregate, that is to say, as a sector of manufacturing with peculiar and distinct links to the polity on the one hand and to other sectors of the
economy on the other. To varying degrees, defence
industry has been shown to have important trade,
technology, employment and spatial effects in those
economies that endeavour to cultivate it. We have
now arrived at the point where it is necessary to
focus on the constituents of defence industry, the
defence firms, or, as we choose to call them, the
military-industrial enterprises (MIEs). The terminology is coined not to add to the plethora of jar-
gon in the defence field, but to make a distinction
between firms that are mildly involved in defence
production and those that are wholeheartedly committed to it. We also carp at the conception of the
'military-industrial firm' proffered by Gorgol,
since that formulation was confined to US firms and,
besides, had no place within it for state-controlled
enterprise.[1] Nor do we subscribe to the notion that
only high defence-dependency in a firm's output warrants its inclusion in the MIE categorisation: the
existence of giant diversified corporations with
relatively low defence-orientation but, none the
less, massive absolute sales of defence items puts
paid to such a simple rationale. Many large, ostensibly civilian-oriented firms should not be allowed
to escape the MIE net. Table 3.1 makes it abundantly
clear that a firm of the stature of GE may acquire
annual defence contracts worth more than $4.5 bil-
lion and yet register only 17 per cent on the
defence-dependency scale. Other conglomerates with
defence sales of almost the same magnitude--United
Technologies (UTC) and Tenneco--are similarly beholden to the defence sector for a minority of sales,
in both instances for just a mite over one-quarter

of turnover. In reality, though, entire divisions of these firms, and companies like them, are devoted to civilian products while others are largely turned over to defence production. In the case of GE, the huge Evendale, Ohio, and Lynn, Massachusetts, plants of the Aircraft Engine Group, the Pittsfield, Massachusetts, site of the Ordnance Systems Division, the Philadelphia centre of the Defense Systems Division and the Syracuse, New York, facility of Military Electronics Systems Operations all rely on defence as their main source of business. For its part, UTC has three divisions which are markedly defence oriented: Norden Systems (avionics) of Long Island, Pratt & Whitney headquartered at Hartford, Connecticut, and Sikorsky Aircraft with its helicopter factory at Stratford, Connecticut. The vast Newport News shipbuilding complex stands testimony to Tenneco's defence interests and, as we have already noted, it is the cornerstone not just of the economy of coastal Virginia but of the entire US naval shipbuilding industry. In short, then, our conception of the MIE encompasses both prime contractors that are highly defence-dependent and the divisions of large conglomerates that are equally circumstanced. Significantly, we include state enterprises as well as those thrown up by the private sector and, as such, allow for the most common form of defence production entity in the world today; namely, the state or semi-state-owned defence enterprise which abounds in Europe and the NICs. With such straightforward terms of reference in mind, we adopt the role of expounders of the characteristics of the MIE in its various manifestations and that task occupies this and the next three chapters. The immediate purpose, undertaken in this chapter, however, is to draw up a schema which highlights the features common to all MIEs and which can act as a point of departure for considering the ramifications of technological and political changes on individual countries and their firms.

USER-SUPPLIER RELATIONSHIPS

User-supplier relationships specific to MIEs depend on a number of factors that are influenced by the technological nature of the product being made, the level of economic development of linked industries, and the whole panoply of institutional arrangements which constitute the polity within which the firms operate. While the details of these factors may vary through time and by place, the essential differences

Table 3.1 : Leading American MIEs

Rank	Firm	(a) 1983 total sales ($ m.)	(b) Fiscal 1983 defence contracts ($ m.)	(b/a). 100
1	General Dynamics	7146	6818	95%
2	McDonnell Douglas	8111	6143	76%
3	Rockwell International	8093[1]	4545	56%
4	General Electric	26797	4518	17%
5	Boeing	11129	4423	40%
6	Lockheed	6490	4006	62%
7	United Technologies	14669	3867	26%
8	Tenneco	14353	3762	26%
9	Howard Hughes Medical Institute	4938	3240	66%
10	Raytheon	5937	2728	46%

Note: 1. Fiscal year-end September 30

Source: Fortune, 30 April 1984.

between the defence environment and the civilian environment remain as constants. Table 3.2 attempts to summarise the full range of differences by apportioning the business environment into those characteristics peculiar to MIEs and those that prevail under market theory, the idealised free-enterprise context regarded by many as being most conducive to civilian economic activity. On the one hand, MIEs are likened to industrial behemoths which are large, coddled, sluggish and definitely antithetical to free-enterprise tenets of perfect competition while, on the other, civilian firms conforming to the principles of market theory are held up as being socially progressive, economically efficient and technologically innovative. Galbraith, for one, has lambasted the MIE for a series of market imperfections:[2]

> the defense contractor is shielded from all vicissitudes that should attend a corporation in a market economy: contracts cover long periods of time; costs are recovered automatically, and a profit is guaranteed; cost overruns are expected, indeed, built into the system by the

67

accepted practice of underbidding to secure a contract.

Such tendencies, it is avowed, distort the economy and have the effect of fossilising military production along lines that are both arcane and inefficient. Certainly, MIEs have experienced in the main great difficulty in coping with civilian markets and this problem extends to their workforces. One observer sums it up in the following way.[3]

> Many people within military industries with skills and experience do not find their training useful in the civilian sector. The skills, and more important, the culture engendered within military industries is highly specific; engineers and managers work in an environment where achieving performance specifications is far more important than producing to cost. The ease with which cost growth is accepted by the buyer....reinforces this style of work. Inefficiency is not only unpunished, it is frequently rewarded. While people experienced in this environment can be retrained, civilian firms have shown a reluctance to recruit such people, reflecting their assessment of the costs.

Strong words indeed, but criticism is not exhausted with the above waspish protestations. In fact, lamentations concerning the labyrinthine nature of government-supplier relationships are legion. As a monopsonist, the government not only sets the product specifications but also the level of demand for the resultant products. It goes further, requiring technical standards of MIEs which are often judged excessive in comparison with the required standards of civilian production and which, in the final analysis, can prove self-defeating. In view of the fact that the technological imperative dominates defence procurement, MIEs are obliged to adopt an 'innovate or perish' attitude in order to stay in the process of bidding for weapons contracts.[4] Yet the government monopsonist requires that these same MIEs be included among the list firms where technical competence has been vouched for through previous defence work or else they are excluded from the bidding process. A paradox is thereby thrown up where for 'newer firms in innovative work, this creates something of a Catch-22 situation....one can't get on the list without prior experience but failure to be on the list is what keeps one from acquiring experience in the first place'.[5] Despite the vaunted aim, then, a monopsonist market structure may have the contradictory outcome of dampening the innovation deemed so vital to defence-industrial competition and weapon effectiveness. The result may be a defence-industrial base composed of a core of MIEs that have difficulty withdrawing from defence production because of their singular inability to adapt to civilian markets, but which, ironically, find themselves protected from competition from newcomers owing to the impediments placed in the way of entry into the defence market by the monopsonist government customer. A change in this cosy situation occurs only when the government paymaster recognises that the existing structure of the MIEs is falling down on its allotted task of advancing technical progress and therefore needs to be shaken up by the introduction of new-technology firms.

The dependence of MIEs on government may be even more pervasive than what is generally the case with American military-industrial firms. At least these last remain privately-owned, structured in the same corporate fashion as other US firms and, with the notable exception of bail-out expedients, markedly devoid of a state presence in their shareholdings. By way of contrast, many MIEs outside the USA are bound to the government for their capitalisation as well as their markets. Such a contingency arises out of the desirability of achieving some form of stability. As elicited from Table 3.2, instability is integral to defence industry. Since defence production conforms to a wave-cycle, more often than not productive capacities of MIEs exceed peacetime equilibrium requirements. At the same time, the gestation period for weapons is lengthy (and getting lengthier) and is a prerequisite before production can come into full effect. Companies unable to simultaneously mount several weapons programmes which fully occupy the workforce must adjust their manpower to fit the differing requirements of a single weapon's life cycle. Thus, labour needs are minimal during the development stages (at least in quantitative terms), at a maximum with series production, and at a moderate level with the long tail end of the weapon's life cycle when maintenance is the chief obligation of the contractor. Loth to lay off labour because of the difficulties involved in reacquiring skilled workers and coping with accusations of social undesirability, MIEs tend to hoard labour despite the penalties thereby incurred of extra cost overheads and reduced profitability. As we have pointed out elsewhere, the circumstance of a strategic necessity for surge capacity combined with

Table 3.2 : MIE Characteristics

Defence	Market Theory (Civilian)
Market Characteristics:	
Structure	
- monopsonistic: dependence on defence procurement	- competitive: sector/industry specific
- Institutionalised inter-dependence between producer and customer (MIC)	- state exclusion from production
- monopolistic/oligopolistic for particular products	- separation of buyer and seller
- substantial barriers to entry and exit	- competition for market shares; importance of advertising
- defence conglomerates	
- competition for entire market, minimal advertising	
Stability	
- contingent upon; (i) state perception of international tension (ii) political factors and state policy	- subject to short, medium, and long-run business cycles; market conditions
- maintenance of R & D, and production capacity for strategic purposes through (i) follow-on system (ii) buy-in (iii) add-ons	- equilibrium forces
- socialisation of risk	
Administration Costs	
- heavy fixed overhead costs	
ratio: <u>Administration Employees</u> 100 production employees Defence = 69.7	Civilian = 36
- Government relations (Information and influence) of paramount importance	
Capacity Utilization	
- low; surge capacity for mobilisation	- higher utilisation rate: 85-95% desired
- excess capacity ranges from greater than 90% (munitions) to 30%	- private-owned plant and equipment
- provision of government-owned plant and equipment in conjunction with private facilities ('Reserve capacity' Requirement)	
Location Constraints	
- political and strategic factors	- market and cost factors (marginal location theory)
- cost/promotion of Inertia	- government influence (e.g. incentives)

Table 3.2(2) : MIE Characteristics (continued)

Internal Characteristics:	
Decision-Making	
- significant government (customer) intervention in decision-making process as to (i) product specifications and standards, and (ii) production process - bureaucratic, organisational and political pressures - production after sale	- management has autonomy over production and marketing decisions - production for inventory
Price	
- prices proportional to total costs:cost-plus contracts, escalation clauses - price factors subordinate to technological factors - negotiated between buyer and seller - "cost and subsidy maximisation" - importance of competence and responsibility	- supply and demand factors (market)
Product Mandate/Research and Development	
- technological criteria dominates product development; trend improvement of a given set of performance criteria - technological dynamism - substantial direct state under-writing of R & D expenses - unique products; particularly on technological criteria	- incorporate economic and technical criteria - product differentiation

a labour situation that conspires to strain profits over part of the weapons life cycle virtually compels some form of socialisation of risk for MIE operations. Such socialisation is frequently manifested through the government underwriting of defence production by means of state ownership. In the AICs, for example, governments have intervened to ensure that defence industry remains available through playing the nationalisation card. The outcome in France was the full nationalisation of the Dassault-Breguet aircraft firm in 1981 and the state's 51 per cent stake in the Matra missile firm.

As for the UK, key MIEs of the likes of Rolls-Royce,
BAe and BS became state enterprises in the 1970s
(although a subsequent British Government had a
change of heart); West Germany's government holding
company, VIAG, ensured that shipbuilding survived;
whereas Sweden nationalised its shipbuilding industry in 1977 before returning (in 1981) the state-
owned part of the aerospace industry--Saab-Scania--
to the private sector as a reward for greater
viability. For ideological reasons, of course, the
Soviet bloc entertains nothing but state-owned MIEs.
Their priority status within the industrial system
is endorsed through a host of privileges: better
machinery and plant than civilian factories and a
workforce enjoying better pay and housing and medical benefits; not to speak of the power given the
managers of defence industry to commandeer what they
need from their less-privileged counterparts in non-
defence plants.[6]

For their part, the NICs offer a wealth of
insights into the operations of state-owned MIEs and
we can do no better than to dwell on their rationale
at this juncture. Almost without exception, MIEs in
NICs are creatures of the state from their very
inception and, indeed, are state enterprises from
the outset. Furthermore, as NIC governments are
enjoined to participate in defence manufacture in
order to save foreign exchange in the first place
and generate linkage effects to stimulate domestic
industry in the second, they find themselves unable
to disengage from direct involvement in local MIEs
because of the inability of these same enterprises
to capture the benefits of market theory. The cyclical nature of defence production requires government
underwriting during production downturns while the
'demanding requirements for quality and precision in
defence-related industries make it difficult for
these concerns to compete effectively in the commercial market' where price and modest technical capabilities reign supreme.[7] Surmounting the day-to-day
economic realities of running MIEs is the wish of
NIC governments to gain the solace of controlling
what are perceived to be sovereignty-enhancing
assets. The poor economic bargain of 'permanent
receivership' of MIEs with its never-ending bill of
state subsidy is countered by the conviction that
the nation is obtaining high technology, triggering
the process of industrial innovation, guaranteeing
weapons supply without external interference, creating jobs--and highly skilled ones at that--while
provoking industrial linkages and clawing back
valuable foreign exchange as previously mentioned.

That these arguments are all subject to valid repu-
diations is frequently overlooked in the nationalist
fervour that grips such ventures.[8] India is salutary
in any overview of state-owned MIEs in NICs. The
country's Ministry of Defence operates through its
Department of Defence Production no fewer than 34 R
& D organisations, 33 ordnance factories and nine
public sector undertakings: together employing
300,000 workers and accounting for up to 15 per cent
of national industrial output. Amongst the portfolio
of products are ships, aircraft, aeroengines, missiles, ground radars, avionics, APCs and MBTs. True
to the Soviet convention, no less than half of the
output of public sector undertakings is destined for
civilian markets (e.g. merchant ships from the Mazagon Dock, Garden Reach and Goa shipbuilding establishments, regional airliners from HAL and telecommunications equipment from Bharat Electronics). As
with the AICs, India cannot escape 'pork-barrel'
influences in the choosing of sites for defence
plants, but unlike them it arrogates to the state
the right to manufacture munitions, aircraft and
ships.[9] The state MIEs are manoeuvred into profitability on the strength of monopoly production
rights and cost-plus contracts. For instance, the
aircraft undertaking--HAL--is allowed to charge its
air force and airline customers the actual cost
incurred in aircraft overhaul and repair plus a
twelve per cent profit margin whereas Mazagon Dock
builds naval minesweepers on a cost-plus basis that
includes an escalation clause to accommodate
increases in labour, materials and overhead costs.
Unsurprisingly, the three key MIEs of Bharat Electronics, Mazagon Dock and HAL were able to achieve
percentage ratios of gross profits to capital
employed of 13.43, 6.98 and 2.80 throughout the ear-
ly-1970s.[10] A decade later, five MIEs remained officially profitable--including HAL and Bharat Elec-
tronics--but the two principal shipyards of Mazagon
Dock and Garden Reach recorded, between them, losses
of R.469 million in 1985-6.

Cost-plus contracts are regarded as essential
to user-supplier relationships in order to amelio-
rate much of the risk involved in developing
advanced weaponry. Since capacity usage is spasmodic
(recall Table 3.2) as a result of wave-cycle induced
production fluctuations, it must be subsidised
through 'reserve capacity' schemes, plant provision
by the state (the GOCO plant or 'agency' factory) or
specially tailored pricing regimes. Subsidising of
capacity is common wherever state-run armouries,
arsenals and dockyards have made the transition from

merely undertaking repair functions to developing the capability to actually make defence items. These MIEs revert to maintenance activities when production orders are in abeyance. In the UK, for example, the Royal Arsenal at Woolwich was the leading artillery R & D and testing establishment as well as the country's main gun-making plant and it could be diverted to the former occupation when peacetime conditions undercut the need for series production of guns. Similarly, the Royal Dockyards were accorded a 'lead' yard role, monitoring the series production of ships in private MIEs through their responsibility for initial ships in each new warship class (e.g. Portsmouth led with dreadnought-style battleships in the early years of this century while, from 1908, Chatham afforded the Admiralty a yardstick for evaluating the cost and technical performance of privately manufactured submarines).[11] The fact remains, however, that private MIEs are also major beneficiaries of state subsidy of 'reserve capacity'. The UK Admiralty saw fit to subsidise the retention of 18,000 tons of armour plate capacity held by three private firms after World War 1 even though demand for capital ships had evaporated and, along with it, any purpose in maintaining the 60,000 tons of capacity built up by five firms prior to the war. In marked contrast to later practice, the USA ensured a 'reserve capacity' capability through its publicly-owned arsenals and dockyards such that: 'naval appropriations generally carried congressional riders, mandating that navy yards or government arsenals produce all material unless excessive costs and time loss result'.[12]

In more recent times, the USA has favoured private MIEs and has socialised their risks through a blend of MIE-biased contracting procedures and the provision of GOCO plants. Cost-plus contracts best exemplify the first and were made legal under the National Defense Act of 1916. Their principle is 'the payment of the full costs by the government, and compensating the contractor for his organisation by the payment of a fee either fixed or in the form of a percentage of the cost'.[13] Initially, war conditions invoked the use of cost-plus contracts in as much as they swept away the peacetime obsession with minimising matériel costs and replaced it with an urgency to mass-produce good-quality defence items. It was felt that the new priority could only be brought about by paying MIEs to boost capacity (and to be subsidised in entirety for so doing) while continuing to invest in the development of new weapons of the highest quality. The latter basis for cost-plus contracts was continued after the war and became more prominent as the technological imperative--and the need to reward firms for innovative work--usurped other concerns of military users. As an added bonus, cost-plus contracts appeared to offer distinct price advantages over fixed-price contracts whenever the possibility arose of extended production runs. The Enfield rifle case was held up as instructive. This British-designed weapon had been made by three private US contractors (Winchester Repeating Arms of New Haven, Remington of Ilion, New York, and Remington of Eddystone, Pennsylvania) for the UK Government under a fixed-price contract which charged $42 for each rifle. Under the cost-plus system subsequently imposed by the US Government, the same basic weapon could be had for only $25 after the attainment of a production run of 100,000. Unfortunately, the post-World War 1 record of cost-plus contracting has abounded with examples of serious cost overruns which have tended to devalue the Enfield example. This has arisen because, in the absence of competition once a MIE has been awarded a production contract, the firm has little incentive to minimise costs: after all, the state is covering the full outlay regardless of delays, snafus and sheer managerial incompetence and is still guaranteeing the firm a profit!

Yet, the alternative of wholesale adoption of fixed-price contracts in lieu of the cost-plus variety is also problematical. During the 1960s the US Navy shifted away from cost-plus to fixed-price contracts as part of its plan to retreat from warship construction in Navy Yards and, instead, to rely increasingly on construction in the more technologically-capable private MIEs. It opted to assign 'lead' ship design and construction contracts to the shipbuilder offering the lowest bid in terms of predetermined delivery time and price. Efficient yards could earn profits by shaving costs below the fixed-price, less-efficient ones were scorched by cost overruns and suffered losses on the contract: in both cases, however, the risk was borne by the supplier and not the customer. In actuality, shipbuilders competed for contracts on the basis of 'buy-ins'(Table 3.2), that is to say, proffering unrealistically low bids for the 'lead' ship (which went undetected by the US Navy owing to its abandonment of design functions and loss of dockyard construction expertise) in the hope of gaining series production contracts for the rest of the class, whereupon profits could be recouped through learning-curve economies. In order to circumvent this

practice, the US Navy resorted to the Total Package
Procurement procedure in 1969 whereby the design and
production of the entire warship class was assigned
at the outset to the lowest bidder. This revamping
backfired, though, as the MIEs continued to underes-
timate costs in their process of bidding for fixed-
price contracts. By 1976 private MIEs were claiming
an extra $2.7 billion from the US Navy as compensa-
tion to accommodate cost overruns in warship construction: a figure amounting to more than half the
US Navy's annual shipbuilding budget.[14] Litton's
Ingalls Shipbuilding Division had overruns amounting
to over $1 billion and it was not alone; Newport
News claimed unanticipated costs of $742 million and
GD's Electric Boat Division claimed a further $544
million.

As a consequence of such situations, the US and
UK governments have moved away from cost-plus and
rigid fixed-price contracts to favour adjustable
contracts which continue to force the MIE to carry
the burden of cost overruns except where the overruns have been reasonably incurred through genuine
development problems. Cost-plus contracts are still
employed, although their usage is restricted to R &
D projects or risky initial production runs of innovative weaponry where, because of the nature of
innovation, cost estimation is largely guesswork.
For their part, GOCO plants also grew out of World
War 1. Their most obvious expression was the provision of twelve fabricating shipyards erected and
owned by the government but operated on the 'agency'
basis by private contractors. Among them was the Hog
Island Shipyard, the largest in the world at the
time. The experience was repeated in World War 2
under the auspices of the government's Defense Plant
Corporation. That body built plants worth $8 bil-
lion which on completion were handed over to private
contractors on five-year leases (at the end of
which, the contractor was given the option of purchase). Supplementing this effort was the provision
of $5 billion of industrial facilities financed by
the US Army and $3 billion by the US Navy to give a
grand total of $16 billion of GOCO plant investment:
a substantially better record than the $9 billion in
plant investment put up by the private MIEs themselves.[15] Many of the government plants remain in
use; for example, GD relies on the Fort Worth GOCO
plant to produce its F-16 fighter while its M1 MBT
derives entirely from GOCO tank plants.

If anything, the socialisation of risk and production has become more pressing in recent times.
This is a reflection of Kaldor's 'baroque'

tendencies, that is, the occurrence of sharply increasing development and production costs with each generation of weapon system in return for relatively marginal advances in technical capabilities. A simple perusal of intergenerational fighter costs is testimony enough: the North American Aviation P-51 Mustang of World War 2 cost $200,000 per copy in current dollars, the company's F-86 Sabre of Korean War vintage was double the price, the Lockheed F-104C Starfighter of the later 1950s was costed at $1.6 million, while the MD F-15 Eagle began life in the early-1970s with a price tag of $9 million but was fetching $45 million a decade later.[16] Real performance improvements occurred with the switch from the piston-engined P-51 to the turbojet F-86 (from 400 mph to 700 mph) while the F-104C effectively doubled the maximum speed of the F-86. However, since the late-1950s combat speeds have remained virtually constant and, instead, technical performance is determined by such criteria as range, weapons load and survivability: properties that are acquired at huge cost as witness the bill for replacing the 1950s-generation fighters (F-101, F-104, F-105, F-106, F-4) with those of the F-15 marque in the first place, and continually upgrading the F-15 model in the second. Evidently, the past twenty years has seen 'baroqueness' set in with a vengeance where fighter aircraft are concerned. All told, in the forty years since World War 2 fighter aircraft have experienced a factor increase in constant prices ranging from 15 to 25 whereas the less 'baroque' MBT has suffered a four-fold increase in unit cost. Current generation strike aircraft are about four times more costly than the machines they are replacing, missiles weigh in at around three-and-a-half times the cost of obsolete systems and newly-commissioning frigates are three times more expensive than their predecessors.[17] It is the drive for technical advances in weapons systems that has overridden market considerations of price competition; added to which is the time factor. Put bluntly, the object is to capitalise on a technical improvement before the opposition can make use of it and, in any event, existing equipment has a finite service life and must be replaced at a predictable time in the future. Technical capability in conjunction with ability to deliver according to schedule are the main formal or official factors in allocating weaponry development contracts to specific MIEs. Moreover, once the MIE has formulated a prototype weapon, it is almost sure to capture the production contract. Not only does it already have the advantage of design and initial tooling facilities, but it is likely to benefit before any would-be competitor from learning-curve economies: a happening which would further embed its preliminary advantage. The only context which would allow alternative suppliers to compete with the original design MIE is one of large production runs where economies of scale are possible for all producing parties. In recent decades, this potential has only been fully realised in the missile sector. It was instigated in 1956 when the US Navy decided to 'second-source' Sidewinder AAM production, choosing to contract GE as a supplementary producer to Philco (later part of Ford Aerospace) and going on to benefit, as a direct result, from a 70 per cent reduction in the cost of missile guidance and control systems between 1956 and 1962.[18] Recent 'second-source' programmes have been the cause of cost savings of 37 per cent between 1982 and 1986 on both the Sidewinder and Sparrow AAM projects--the former now being produced by Raytheon as well as Ford and the latter emanating from Raytheon with GD as second source.[19] A cautionary note needs sounding, however, for not all 'second-source' contracts escape start-up problems. For example, the attempt of the DoD to compel Pratt & Whitney to 'second-source' GE's F404 fighter turbofan engine has not proceeded smoothly; indeed, 'identically shaped pieces for....Pratt & Whitney-built engines differ enough from General Electric's to amount to new parts, pushing up costs beyond initial forecasts'.[20]

Nevertheless, the 'second-source' experiment has been sufficiently encouraging to move governments in the direction of competitive tendering. The evidence points to the lack of economies of scale in the production of warships and ammunition but the prospect of them in the manufacturing of aircraft, missiles and armoured vehicles; and it is in those latter areas that governments have been most attentive to the question of alternative suppliers.[21] The UK, for example, has taken the fairly radical step of amending its Defence Contract Condition 15 so that prototype builders are no longer entitled as a matter of course to receiving the first production contracts: rather, the MIE responsible for design and development work must now compete with any other qualified MIE for production work. As a result of this change of heart on the part of the user, the value of competitive contracts awarded by the MoD rose from 38 per cent of all defence contracts in 1983-4 to an estimated 60 per cent in 1985-6.[22] However, the fundamental stumbling block has not been

surmounted; namely, how to achieve the goal of scale
economies across a range of defence products when
numerous circumstances militate against its accom-
plishment. It behoves us to recollect that even in
World War 2 British production runs of aircraft
scarcely merited scale economies owing to the fre-
quent changes in model specifications. It was concluded that a series run of at least 1,500 aeroplanes in which specifications were 'frozen' was the
absolute minimum necessary for such economies;
indeed, runs of 500 or less were more economically
made using bench tools rather than the jigs that are
indispensable in large-scale output.[23] Of course,
the shrinkage of production orders that has been
endemic since World War 2 in a period of peacetime
equilibrium has not only served to make the realisation of scale economies much more elusive, but has
also contributed significantly to the rising intergenerational costs of weapons.[24]

All AIC governments have urged their MIEs to
chase after export opportunities in order to boost
production runs. France maintains that its aircraft
industry must export a minimum of 50 per cent of its
entire output. For example, during the 1974-8
period, no fewer than 297 of the 338 Mirage fighters
ordered from Dassault-Breguet were for export, as
were 272 of the 318 helicopter orders received by
Aérospatiale, more than 50 per cent of that firm's
anti-tank missiles and 75 per cent of Matra's missiles. Nor was the vehicles sector overlooked: 73 of
the 103 AMX-30 tanks ordered from Atelier de Construction Roanne during those years were for foreign
customers. While less pronounced than France,
exports also take up a significant proportion of UK
defence output; figures in the order of 25-30 per
cent have been suggested.[25] In addition to reaching
out for scale economies, exports also help governments aspire to their social objective of ensuring a
high degree of employment stability in defence
industries and, in so doing, fall in line with the
wishes of European MIEs which prefer to hoard
labour. Exports are not invariably forthcoming,
however, and in the modern era of collaborative programmes and co-production arrangements they no
longer constitute an easy way to bolster production
runs. Consequently, governments have prevaricated
where scale economies are concerned and, instead,
have pressed the MIEs to gain learning-curve economies. Like scale economies, they too are a function
of total quantity produced of a given defence item.
Unlike scale economies, however, they are a function
of cumulative output. Economies of scale are

determined by the rate of production for a specified
period of time and stem from the fact that an
increasing tempo of output makes feasible the use of
specialised plant and labour. In comparison, learning-curve economies derive from the experience
acquired by the workforce and management over the
duration of the production contract: efficient work
practices discovered in the early phases are incorporated into subsequent shop-floor activities to the
betterment in building time and cost of the later
units and to a reduction in programme costs as a
whole.[26] Empirical studies indicate that labour
learning-curves of 80 per cent and 75 per cent are
typical in the UK and US aircraft industries respectively. In terms of economies, the first figure
translates to a reduction of unit aircraft costs by
ten per cent for each doubling in cumulative output.[27] The recently concluded USAF AGM-86B air-
launched cruise missile programme offers a concrete
example of the realisation of learning-curve economies. Production time for each missile was cut from
5,500 hours to only 880 hours by the 1,000th unit
with commensurate cost savings over the full
1,715-missile production run.[28]

The Bail-Out Expedient

In the final analysis, governments have the option
of sustaining their defence-industrial base regardless of the profitability--or otherwise--of the
individual MIEs from which it is constituted. This
'permanent receivership' is costly for the public
purse and may be spurned for the individual firm if
the government is satisfied that the industrial base
contains alternative suppliers. Apart from state
acquisition of financially-troubled MIEs, the government can intervene to protect a MIE by affording
it contracts at judicious intervals--the follow-on
system--or by bailing out the firm through a capital
injection or an enforced restructuring within the
bounds of the private sector. This latter strategy
includes the forced merger of several MIEs at government instigation.[29] As an aside, it is worth noting that the bail-out expedient sometimes owes its
inspiration to geographical considerations and we
shall give full play to this aspect of 'permanent
receivership' below. To begin with, however, we dis-
cuss the follow-on system which was given the appellation of an 'imperative' by Kurth, its publicist.
    The follow-on system is the simplest tool in
the 'permanent receivership' toolbox in that it

retains the existing ownership and structural characteristics of the MIE. It finds favour with governments precisely because it is in keeping with a minimal interventionist stance on their part. In
essence, the government is charged with finding
replacement programmes for existing ones so that the
new project can be inserted into the production line
on the termination of the old system. According to
Kurth, an aerospace MIE is entitled to a new project
about three years prior to phasing out of production
of the existing one if the follow-on imperative is
operating true to form.[30] For the general case, Gansler puts it about that a defence contractor needs
from seven to ten years to develop a new weapon system and a succeeding period of three to five years
in order to produce it.[31] Moreover, for the follow-
on system to be most effective it must reinforce
specialisation among MIEs. This is vital as the
smooth progression of production from one system to
the next is facilitated when the new project is
essentially of the same technical mould as the old
one, which is to say, builders of fighters should
receive contracts for successor fighters, builders
of SAM missiles should receive the contracts for
their replacements and constructors of last-year's
nuclear submarine should follow with this year's
boat and so on.

Claiming to have detected the follow-on imperative as underlying most US weapons programmes of the
1960s and 1970s, Kurth discloses some interesting
examples of how MIE viability is bolstered by government development and production contracts. An
attempt to both summarise and update them follows.

GD.
In 1960 GD was confirmed as a bomber specialist with
production of the B-58. This aircraft was phased out
in 1962 and replaced by development of the F-111
which, despite its designation, was primarily a
bomber. Full-scale production of the F-111 extended
from 1966 to 1974 when--in place of a new bomber
requirement--the company was given a development
contract for a lightweight fighter which subsequently materialised as the stunningly successful
F-16.

Rockwell International.
As North American Aviation, this company worked concurrently on development of the B-70 bomber and
Apollo space programme in the early-1960s. Cancellation of the bomber was offset by production contracts for Apollo, contracts which maintained

Rockwell's aerospace interests until 1972 when
development work on a new bomber, the B-1, and the
Shuttle space vehicle replaced it. Cancellation and
reinstatement marked the B-1, but it and the Shuttle
sufficed to underpin Rockwell into the late-1980s.

Boeing.
Boeing entered the 1960s with a running down of B-52
bomber production in favour of a build up of Minuteman ICBM production. Successive improvements of
Minuteman kept production lines busy throughout the
1960s and 1970s until cruise missile and AWACS programmes replaced them.

MD.
The fighter 'house', McDonnell, experienced a long
production run of F-4s throughout the 1960s until
1972, whereupon MD's F-15 successor was readied for
production and has kept the MIE busy ever since.
From the late-1970s, a stable mate, the F/A-18, has
also issued from St Louis. Douglas was forced into
merger with McDonnell partly on account of its faltering missile programmes (i.e. the cancellation of
Skybolt). In the 1970s, defence contracts returned
again to Douglas, initially with the C-9 transport
(a military version of the DC-9 jetliner), then the
KC-10 (a tanker variant of the DC-10 widebody airliner) and, as a follow-on project into the 1990s,
the C-17.

Grumman.
A US Navy supplier, Grumman was kept going throughout the 1960s largely on the strength of NASA and
USAF (F-111) subcontracts as well as steady orders
for US Navy A-6 strike and E-2 AEW aircraft. These
were supplemented in the 1970s by the F-14 fighter
as the subcontracts were run down. Yet, the company
went through a precarious period in the 1970s when
its survival was put in jeopardy.

Lockheed Missile & Space Company.
This division of Lockheed entered the 1960s with the
Polaris SLBM programme in full swing. In turn, this
was first replaced by Poseidon and then by the current-generation Trident SLBMs.

Lockheed-Georgia Company.
Building the highly successful C-130 turboprop
transport at the beginning of the 1960s, this division introduced the C-141 jet transport into production in 1964 and replaced it with the even bigger
C-5A transport in 1968. By the 1980s, an upgraded

model (the C-5B) was returned to production, the
lull having been occupied by on-going production of
C-130s.

LTV.
LTV started the 1960s with the F-8 fighter in full
production, a state of affairs which continued until
1966 when the A-7 strike aircraft superseded the
fighter on the production lines. A-7 output was
maintained through to the mid-1980s, latterly as a
result of 'add-ons'.

These examples clearly confirm the thesis of a
follow-on system, at least in so far as the aforementioned MIEs are concerned. However, two codicils
need stressing. In the first place, the evidence
does not wholeheartedly support the view that MIE
product specialisation will be consolidated as a
consequence of follow-on ventures. Certainly, both
of the Lockheed companies as well as McDonnell and
Grumman conform to the assertion, but all the others
are distinctly at variance with it. GD was obliged
to forsake bombers for fighters in the 1970s, Rockwell was compelled to alternate between bomber and
space vehicle projects, Boeing made the transition
from bomber to missile work before diversifying yet
again through its development of AWACS programmes,
Grumman leavened its fighter offerings with aircraft
of the AEW and strike variety, while LTV moved from
fighter to strike aircraft. The last MIE is also
illustrative of the second codicil; namely, the
extension of production runs purely as a result of
political lobbying and despite the wishes of the
armed forces. Known as 'add-ons', these extensions
are supplementary allocations for extra production
contracts introduced into the defence budget as it
proceeds through the legislative approval process.
At that juncture, legislators with a vested interest
in ensuring continued activity in MIE plants located
in their constituencies can insist upon such mili-
tarily-redundant 'add-ons' in return for concurrence
with large programmes dear to the hearts of other
legislators and the military. The Texas delegation
to Congress was renowned for its ability to perpetuate the LTV A-7 production run long after military
interest in the aircraft waned (it was relegated to
the Air National Guard, a reserve formation) and, in
so doing, effected high activity levels at the Dallas plant.[32] Needless to say, application of the
follow-on system does not appear to be confined to
the USA. In Western Europe, the recent extension of
the Panavia Tornado aircraft programme smacks of

83

this system. A product of BAe, MBB and Aeritalia,
Tornado was initially planned to go out of production in 1989 on the completion of 809 aircraft for
the British, West German and Italian services. Yet,
belated export orders for 80 machines (72 for Saudi
Arabia and eight for Oman) when combined with an
extra batch of 35 for the Luftwaffe and nine for the
RAF have extended total orders and pushed back the
production completion date into the 1990s.[33] This
programme extension is important because it cuts the
'down' time wherein the three MIEs are devoid of
major combat aircraft production before the follow-
on Eurofighter can be brought on stream. If Panavia
--the holding company for the Tornado's participant
MIEs--is successful in achieving more export or 'at-
trition' orders, then, the gap between Tornado
phase-out and Eurofighter phase-in may be eliminated
altogether.

Notwithstanding the clear confirmations of the
follow-on system, nagging doubts persist as to its
general applicability. For one thing, the system
appears selective in disbursing favours: for example, after a highly profitable period in the 1960s
and 1970s selling F-5 fighters to US client states
with the help of the US Government (courtesy of Foreign Military Sales credits), the Northrop Corporation was unable to reap the benefits of official
sanction for its F-20 follow-up aircraft in spite of
expending $1.2 billion on the machine's development.
This is not to say that Northrop was overlooked by
government: on the contrary, its development award
of the Advanced Technology Bomber intended as a successor to Rockwell's B-1B in the 1990s vouched for
its future, as indeed did its selection as one of
the two winners of the Advanced Tactical Fighter
prototype contract which came at the same time as
the DoD formally scotched the F-20's chances of
becoming part of the US inventory. Provisionally,
the prototype contract should create work amounting
to 1,000 jobs while the failure to slot the F-20
into the void left by the F-5 should cut about 2,000
jobs: the net outcome being a job loss of 1,000 at
the Hawthorne, California, plant.[34] The case seems
to indicate that, first, 'private venture' R & D,
such as the F-20, is horrendously risky and that,
secondly, virtually any weapon system design outside
the purview of the government monopsonist is
scarcely worth contemplating. The fact remains, however, that Northrop eventually avoided corporate
failure owing to government support for other programmes: in short, it ultimately benefited from
'permanent receivership'.

84

Other MIEs were not so fortunate. The treatment meted out to some firms appeared to directly contradict the follow-on system. In the US alone, airframe makers Curtiss-Wright, Douglas, Fairchild, Martin and Republic all failed to receive badly-needed follow-on contracts.[35] Indeed, the urgency of the need was so acute for some of these MIEs that government indifference during the crucial hiatus in their activities was tantamount to sounding their death knells. Thus, cancellation of the 1948-vintage XF-87 four-jet fighter for the USAF ended Curtiss-Wright's airframe ambitions and forced it to confine its aeronautical interests to the engine field, while failure to entice the same service to commission a follow-on design to the F-105 tactical fighter cost Republic its corporate independence in 1965 (it was absorbed by Fairchild). The reality whereby these firms contained both spare capacity and expertise was not instrumental in attracting government support: a reaction which seems to fly in the face of the declared aim of the follow-on system, that of maintaining reserve capacity. Therefore, the only conclusion to be drawn is that a surfeit of capacity was available in the defence niche occupied by those MIEs--by and large the fighter and attack aircraft arena--where alternative suppliers were deemed better able to meet the requirements of the DoD. The apparent unconcern expressed by the government at the plight of Grumman in the 1970s perhaps bears witness to this rationale. Financially strapped on account of the cost of developing the F-14 Tomcat fighter for the US Navy, the firm challenged the government to renegotiate its fixed-price contract for supplying production aircraft in 1972. Essentially, Grumman refused to complete a batch of 48 aeroplanes unless the price was increased as it claimed a loss of $135 million in building 86 earlier machines. Prolonged disputation likely soured relations between Grumman and the DoD although, eventually, the government was forthcoming with an additional $806 million to procure the 48 aircraft in question. In the meantime, however, Grumman's finances were in such desperate straits that it tottered on the verge of collapse. It was rescued, in fact, by an Iranian order for 80 F-14s accompanied by a $200 million loan arranged jointly by the Melli Bank of Iran and sundry US banks, and it only escaped bankruptcy by the advance payment of $28 million mediated by the Shah.[36] So far as the DoD was concerned, alternative supplies of naval fighter and attack aircraft could be obtained in a pinch from MD and LTV.

Grumman notwithstanding, government intervention is often regarded as a last resort measure to prevent the dissolution of productive assets. Such was perceived to be the case in 1971 when the US Government stepped in to underwrite Lockheed while the UK Government actually nationalised Rolls-Royce partly owing to the belief that both firms were too vital to the Western defence effort to be allowed to go under. A similar rationale had underscored an earlier UK Government intervention in the aircraft industry, that concerning the de Havilland firm in the mid-1950s. As a consequence of design flaws in the Comet civil jetliner, this enterprise was left with £15 million worth of unsaleable aircraft and useless jigs and tools. Rather than see the demise of a firm with important military expertise, the Treasury ordered modified Comets for the RAF and poured aid to the tune of £10 million into de Havilland which proved sufficient to tide it over its difficulties.[37] A comparable background surrounded the 1974 decision of the UK Government to bail out the Ferranti electronics manufacturer. In return for a £15 million injection and a 62.5 per cent equity stake, the state guaranteed the survival of a MIE holding valuable technical capabilities in the defence field.[38] Contemporary examples of bail-outs are not difficult to find. Creusot-Loire had its severe financial difficulties of the early-1980s waived as a result of the desire of the French Government to maintain the firm's 'position as the sole French manufacturer of main warship armour'.[39] Refits of armed patrol vessels for the US Coast Guard arrived in the nick of time to keep in being the Seattle ship-repair facilities of Todd Shipyards Corporation in 1985-6. In an effort to sustain the De Schelde yard at Flushing, the Netherlands Government took the unusual step in 1984 of ordering four M-class frigates long before the ship design had been completed.[40] A decision made in 1986 by the USAF to spend $180 million on the purchase of 80 C-21A staff transport aircraft that had been leased to it by Gates Learjet Corporation not only removed the loss-making contract from Gates's books, but saved the company from the painful duty of retrenching in Arizona and Kansas--a contingency which played no small part in drumming up congressional approval for the USAF action.[41] In order to emphasise its new-found reliance on the military, the Gates Rubber Company, acknowledging its own limitations as a defence contractor, was anxious to sell its erstwhile commercial business jet subsidiary to a reputable MIE.[42]

These are all instances of 'bail-outs' undertaken in support of the existing corporate entity;
but state support may be predicated on a drastic
shift in corporate structure. The Cosmos case is
illuminating in this respect. The Cosmos firm of
Bristol, England, developed an outstanding military
aeroengine--the Jupiter--but the engine's evolution
was foreclosed by the demise of the parent group in
1920 following rash speculations in Russia as well
as cash advances to the engine department which were
unsustainable. Desirous of procuring the developed
version of the Jupiter but unwilling to restore Cosmos, the UK Government urged the neighbouring Bristol Aeroplane Company to acquire the aeroengine
assets of Cosmos and held out the prospect of aid
provided Bristol committed itself to further develop
the engine. For £15,000 Bristol obtained assets
worth £60,000--plus first-rate engineers--and went
on to receive a government contract for ten experimental Jupiters worth £25,000 which was followed, in
short order, by a production order for 42 engines.[43]
Thus, a useful piece of technological capital was
saved for the nation even though its innovating MIE
was not. Also, the fact must be recognised that
bail-out measures sometimes turn out to be ineffec-
tual. In 1986, for example, the Defence and Commercial Craft Division of Watercraft Ltd was closed
down along with its Shoreham shipyard despite having
just produced the first of a series of 14
P2000-class patrol boats for the MoD. Evidently,
the generous awarding of contracts by the state that
were intended to sustain the firm could not overcome
the enterprise's structural problems.[44] By the same
token, orders worth £8 million for two River-class
minesweepers were placed by the MoD with the Cle-
lands shipyard at Newcastle in September 1983 for
the express purpose of providing jobs in an area of
high unemployment and, incidentally, were intended
to give the yard enough work to ensure its survival.
However, subsequent rationalisation undertaken by
the BS parent firm led to the expunging of the yard
and the reallocation of the minesweeper contracts.[45]

The Locational Side of Bail-Outs

This last case introduces a further reason for the
application of bail-out expedients, that is, to save
jobs in politically-sensitive locations. Of course,
this rationale is implicit in pork-barrel politics
in any event, but its specific application in support of failing MIEs is worthy of mention. By all

accounts, it has long antecedents in the UK where
the use of naval contracts to eradicate real or
impending unemployment in communities dependent on
MIEs can be traced back at least as far as the late
Victorian era. The conversion to bail out seems to
have come about surreptitiously, as is implied from
the following comments.[46]

> As recently as 1869, Gladstone's government had
> summarily closed Woolwich Dockyard under its
> programme of 'Peace and Retrenchment' but had
> not considered it had any duty to the thousand
> employees...By the 1880s, government was being
> called upon to use its powers to grant con-
> tracts to relieve distress.

One manifestation of the use of naval contracts to
this end was the sustenance of a major shipyard in
London (the Thames Iron Works at Blackwall) long
after merchant shipowners had forsaken the Thames
for more cost-competitive yards in northern England,
Belfast and the Clyde estuary. Indeed, the Thames
Iron Works even persuaded the Admiralty to order
from it a super-dreadnought and managed to survive
until 1912 and the battleship's completion. In its
latter years the firm's cost uncompetitiveness was
well-known (e.g. its 1903 tender for a steel training ship was £92,800 with a delivery time of 15
months--markedly inferior to the winning bid of
£55,525 and twelve-months delivery time submitted by
Vickers) and speculation was rife that Admiralty
orders coming its way were nothing more than
instances of blatant favouritism.[47] Government
underpinning was finally withdrawn when the firm's
costs became intolerable and it became apparent that
veiled subsidies could no longer be justified with
the availability elsewhere of surplus capacity among
more efficient shipbuilders.

Unemployment considerations have continued to
loom large in influencing the allocation of UK and
other nations' naval contracts. They are, in fact,
part and parcel of the process employed to nourish
hard-pressed shipbuilding companies.[48] The same can
be said for state intervention in the aerospace
industry. Major programmes and their controlling
MIEs have been maintained by the UK Government at
least in part as a consequence of the desire to
obviate localised unemployment (including the bail-
out of Rolls-Royce).[49] As far as France is concerned, the question of job support is a major moti-
vating force accounting for the unstinting backing
given by the government to all the country's MIEs.

Because of their wide-ranging dispersal throughout
France, these firms generate 'a broad base of interest and potential support for the arms industry and
pressure for a strong export policy' which contrive
to officially sanction moves that indirectly mitigate local unemployment.[50] Perhaps the most extreme
case in France of dispersal of work to revitalise
specific aircraft facilities and their host communities was the early postwar flying-boat project, the
Latécoère 631. Government orders for eleven aircraft were divided between Latécoère's own plant at
Toulouse (four) and state enterprise Sud-Ouest (four
from the St Nazaire plant and three from Le Havre)
with six hulls subcontracted to another state enterprise, Nord, having factories in the Paris basin.
Interestingly, the flying-boat proved unsuccessful
in service and, as feasible follow-on designs
appeared to be unnecessary, Latécoère abandoned flying-boat manufacture for missile development.[51]

Even the USA pays lip service to the notion
that defence contracts may serve a useful function
in ameliorating unemployment distress. Dating from
the Korean War, the US Government articulated its
intention of channelling contracts into 'surplus
labour areas' through the issuance of Defense Manpower Order No 4. However, the intention was stillborn as, from 1954, Congress prohibited the purchase
of defence items from contractors located in
depressed regions if they were unable to meet the
lowest available competitive price. After 1960 the
government changed its tack, encouraging successful
MIEs either to locate a portion of their capacity in
depressed regions or, in lieu of that, to place work
with subcontractors already located in them. Heightened awareness of local distress conveyed to Congress was responsible for the 1980 decision of US
legislators to allow the Defense Logistics Agency to
grant $3.4 billion in defence contracts to areas of
high unemployment so long as the cost penalty over
lower bidders in more prosperous areas did not
exceed five per cent.[52] On the whole, though, the
discernible results of this tentative form of US
regional policy were judged to be fairly inconsequential.[53] Moreover, there is enough circumstantial
evidence to suggest that US firms may attempt to
circumvent labour problems through transfer of production facilities to communities characterised by
workers that are either less organised than those at
the original site or more willing to accept company-
imposed work conditions. Conceivably, increased
unemployment at the original site could be a by-
product of such action. The aeroengine makers fall

into this category. They have adopted 'multiple-sourcing', or the placing of subcontracts with more than one vendor, in order to ensure that parts are delivered on time and not held up as a result of labour disruptions. Similarly, they have opted for investment in greenfield sites with nonunionised labour traditions, the so-called 'parallel-production' facilities--duplicate capacity--so as to be able to reduce the bargaining power of labour at the original, unionised sites. In this vein, GE erected a plant at Ludlow, Vermont, in the late-1950s and repeated the exercise a generation later with new plants at Rutland in the same state and Madisonville in Kentucky. For its blades and vanes supply, Pratt & Whitney can rely on the newer non-unionised North Berwick, Maine, plant to act as a back-up to the older, unionised North Haven, Connecticut, plant.[54] When all is said and done, however, US depressed regions can resort to the same tactic as their better-off counterparts; that is to say, they can undertake lobbying in order to gain from 'pork-barrel' contracts. Evidence for the efficacy of pork-barrelling initiatives is overwhelming, as instanced above, but its use in averting impending calamitous unemployment is less demonstrable. Certainly, the persistent attempts throughout 1986 of the New York delegation in Congress to rescue the T-46 programme in the face of USAF hostility must be seen against the backdrop of a dire alternative involving the closure of Fairchild's Farmingdale, Long Island, plant and the dispersion of its workforce (in the event, the efforts proved futile: the T-46 was finally cancelled in 1987). Ironically, the same plant was the venue for the 1965 collapse of Republic owing to an earlier bout of USAF indifference, albeit the facilities were retained and eventually put to use making the A-10 attack aircraft under Fairchild's tutelage.

THE DRIVE TO ACHIEVE MIE BIGNESS

While it is patently obvious that MIEs function differently from their commercial counterparts anchored to market theory, it is equally apparent that they do share many of the organisational forms of the second and larger group. Like manufacturing enterprises as a whole, MIEs are driven by the same impulses for growth which have the outcome of enhanced industrial concentration. And like commercial firms everywhere, they are subject to the dictates of technological change, which is to say,

their products succumb to a convoluted evolutionary
process that takes them from the stage of innovation
to that of obsolescence, a process with important
implications for each enterprise in terms of its
competitive position and geographic location. Yet,
MIEs differ in fundamental ways from civil firms and
this makes the drive to bigness and augmented technological capability loom even larger in their corporate game plan. In the first place, MIEs must confront the dilemma of a record of low profit
achievement in relation to the huge costs entailed
in developing weapons systems: a much more pressing
problem than any generally confronting producers in
civil markets. Secondly, the level of uncertainty is
heightened by virtue of the monopsony defence market
on the one hand along with the technological imperative on the other.[55] MIEs are not only government
clients but are hostage to technical dynamics to a
far greater extent than most civil firms owing to
their dedication to performance criteria rather than
cost considerations. In practice, this means that
the expending of large sums on weaponry development
does not always lead to the MIE gaining a production
contract. Thirdly, orientation to serving the government customer tends to 'distort' the MIE in a
fashion hardly commensurate with attaining the ver-
satility required to compete successfully in civil
markets and, in consequence, has the effect of
'locking in' the MIE to defence demand.

Industrial Concentration

Excess capacity, a top-heavy administration geared
to government contract bidding, and the possible
outcome of losing the bid in any event, all conspire
to push the MIE in the direction of acquiring the
supposed security that flows from boosted size. In
the USA, for example, the defence industry--regard-
less of its constituent sectors--is definitely a
highly concentrated industry in that the largest
eight firms produce at least half of the industry's
output. Some branches of defence production--sur-
veillance and detection satellites, space boosters,
fighters and attack aircraft--have fully 100 per
cent of defence contracts monopolised by the leading
eight firms of each branch. Most of the others have
the top eight firms capturing 90 per cent or more of
the contracts. Important exceptions to the high levels of concentration were evident only in the area
of shipbuilding (i.e. warship construction and
marine engineering), drones and ECM.[56]

Table 3.3 : Concentration in US Naval Construction

Warship capability	1945	1955	1965	1985
Aircraft Carriers[1]	Bethlehem (Quincy) Newport News New York NY New York SB Norfolk NY Philadelphia NY Todd-Pacific (Tacoma)	Newport News New York NY New York SB	Newport News	Newport News
Cruisers	Bethlehem (Quincy) Bethlehem (San Fransisco) Cramp Federal Newport News New York SB Philadelphia NY	Bethlehem (Quincy)	Bath Iron Works Mare Island NY New York SB Puget Sound NY Todd (San Pedro)	Bath Iron Works Litton
Destroyers[2]	Bath Iron Works Bethlehem (Quincy) Bethlehem (San Fransisco) Bethlehem (San Pedro) Bethlehem (Staten Island) Consolidated Federal Todd-Pacific (Seattle)	Bath Iron Works Bethlehem (Quincy)	Puget Sound NY Todd (Seattle)	Litton
Submarines	Boston NY Cramp Electric Boat Manitowoc Mare Island NY Portsmouth NY	Electric Boat Mare Island NY Portsmouth NY	Electric Boat GD (Quincy) Ingalls Mare Island NY Newport News New York SB Portsmouth NY	Electric Boat Newport News

Notes: 1. excludes helicopter carriers
        2. excludes destroyer escorts (frigates)

Even in those branches of defence production, however, the drive towards concentration is inexorable. As can be elicited from Table 3.3, the number of prime contractors occupied in producing major warship types has diminished dramatically since World War 2: a sure indication of the urge to concentrate. The seven MIEs actively engaged in aircraft carrier construction in 1945 had been cut to three a decade later and by the 1960s there was but a single survivor; namely, Newport News. Cruiser builders had been whittled down from seven in 1945 to just two in 1985, submarine specialists were curtailed to the extent of witnessing a drop in numbers from six to two in the same time span, while destroyer builders suffered the greatest attenuation with the eight yards of 1945 being reduced to one by 1985. Not to be outdone, other NATO nations have followed suit (Table 3.4). As the archetypal surface warship of the modern age, the frigate has been promoted by the navies of most powers and has benefited from new construction programmes. As a corollary—and somewhat paradoxically—the number of MIEs bidding for frigate orders has steadily diminished. Thus, Canada's half-a-dozen yards of the 1950s and 1960s had been reduced to two by the 1980s, France's four producers of the 1950s had halved in the 1960s and halved again in the decade following, while the UK's 15 frigate builders of the 1960s had shrunk to a mere two survivors a generation later. Only Italy bucked the trend, retaining the same number of yards (two) in the 1980s as it had in the 1950s although, even here, the ownership had been concentrated into the hands of a monopoly supplier: the state's Fincantieri group.

Similar trends are discernible throughout the range of defence production. Table 3.5 singles out MBTs from the array of land weaponry and charts the evolution of the US Army's principal types since World War 2. The chief American tank of that war, the M4 Sherman, was built by ten producers in the USA and one in Canada. For its part, the Korean War-vintage M47 issued from only two plants, although its Cold War successor—the M48—was built in sufficient numbers to warrant output from four tank plants. By the 1960s, the follow-on M60 was rolling out of two plants (later reduced to one) under the control of a single organisation; namely, Chrysler. The current-generation M1 is also the responsibility of only one MIE—although this is now GD and not Chrysler—while acceding to the principle of second-sourcing ensures that a tank production base comprising two assembly plants is perpetuated.

Table 3.4 : Rationalisation of NATO Frigate Producers

Country	Years: 1950-55	1960-65	1970-75	1980-85
Canada	Burrard Canadian Vickers Davie Halifax SY Marine Industries Victoria Machinery	Burrard Canadian Vickers Davie Halifax SY Marine Industries Victoria Machinery	Davie Marine Industries	Marine Industries St. John SB
France	Arsenal de Lorient AC de la Loire FC de la Mediterranée Penhoët	Arsenal de Lorient AC de Nantes[1]	Arsenal de Lorient	Arsenal de Lorient
Italy	Ansaldo CNT	CNR, Riva Trigoso CRDA Castellammare	CNR, Muggiano CNR, Riva Trigoso	CNR, Muggiano CNR, Riva Trigoso
UK	Cammell Laird Denny Devonport DY Fairfield John Brown Harland & Wolff Hawthorn Leslie Portsmouth DY Scotts Stephen Swan Hunter Thornycroft Vickers, Tyne White Yarrow	Cammell Laird Devonport DY Fairfield John Brown Harland & Wolff Hawthorn Leslie Portsmouth DY Scotts Stephen Swan Hunter Thornycroft Vickers, Barrow Vickers, Tyne White Yarrow	Vosper Thornycroft Yarrow	Swan Hunter Yarrow

Note: 1. Built for export only

Table 3.5 : Reduction of US Producers of MBTs

Tank (production period)	Producers
M4 (1942–5)	American Locomotive Co Baldwin Locomotive Works Chrysler (Detroit) Federal Machine and Welding GM (Fisher Tank Div) Ford Lima Locomotive Works Montreal Locomotive Works (Canada) Pacific Car and Foundry Pressed Steel Car Pullman Standard Car
M47 (1949–54)	American Locomotive Co Chrysler (Detroit)
M48 (1952–59)	Chrysler (Delaware) Ford GM (Fisher) Alco Products
M60 (1959–85)	Chrysler (Delaware) Detroit Tank Factory (Chrysler, then GD)
M1 (1982 to date)	GD (Detroit) GD (Lima)

Source: Derived from Jane's Armour and Artillery 1985-86 (Jane's Publishing, London, 1985)

Elsewhere, tank production also happens to be confined to one or, at most, very few MIEs in each country. As evinced through Table 3.6, the European AICs are in a comparable position to the USA (although West Germany has three firms engaged in tank design and production; a reflection of the Wehrmacht's historical legacy). They are, equally, in a comparable situation to the NICs in so far as concentration is concerned. Entering the tank production field in recent years, the latter have fostered state-supported enterprises to replace tank imports from the AICs and, with the noteworthy exception of Brazil, are equipped to sustain only one such enterprise per polity.

Table 3.6 : Tank Manufacturing MIEs

AICs	NICs

AICs	NICs
France: ARE	Argentina: TAMSE (late 1970s)
West Germany: Krauss-Maffei	Brazil: ENGESA (early 1980s)
Krupp Mak	Bernardini (late 1970s)
Thyssen Henschel	Egypt: Factory 200 (late 1980s)
Italy: OTO Melara	China: state arsenals (from 1957)
Switzerland: Contraves	India: Avadi Company (from 1965)
UK: RO Leeds	Israel: state industries (from 1974)
Vickers	North Korea: state industries (since 1970s)
USA: GD	South Korea: start-up in late 1980s
Teledyne Continental	

Source: Derived from listing of MBTs and medium tanks in Jane's Armour and Artillery 1985-86. (Jane's Publishing, London, 1985).

Aeroengine manufacture can also be seen to fit the mould. The predominant mode of aircraft propulsion in 1945 was the piston engine and no less than ten US firms partook of its manufacture (Table 3.7). Ten years later the number had halved--partly, it must be said, as a consequence of technological obsolescence in the product--and by the 1960s only two firms entertained any piston engine development and production. By way of contrast, however, the number of participants in turboprop (slow, propeller-driving turbines) and turboshaft (helicopter turbines) production has remained constant over the four decades (although the constituent firms have varied) whereas the number of turbojet and turbofan manufacturers has actually increased. All told, and despite a change-over in product technology from piston to turbine engines, the trend is still one of concentration: the 16 separate enterprises which undertook some aeroengine production in 1945 had been cut back to seven by the mid-1980s. On a wider

96

Table 3.7 : Concentration in US Aeroengine Supply

Engine type	1945	1955	1965	1985
Piston	Allison Chrysler Continental[1] Franklin Jacobs Kinner Lycoming Packard Pratt & Whitney Wright	Continental Franklin Lycoming Pratt & Whitney Wright	Continental Lycoming	Continental Lycoming
Turboprop or Turboshaft	Allison Boeing GE Northrop Pratt & Whitney	Allison Boeing GE Lycoming Wright	Allison Boeing Garrett GE Lycoming	Allison Garrett GE Lycoming Pratt & Whitney
Turbojet or Turbofan	Allis-Chalmers Allison GE Menasco Westinghouse	Allison Continental Fairchild GE Pratt & Whitney Westinghouse Wright	Continental GE Pratt & Whitney	Continental Garrett GE Lycoming Pratt & Whitney Williams

Note: 1. Renamed Teledyne CAE after 1969.

scale, the tendency was for AICs to rationalise their aeroengine assets. For example, the UK merged all its significant producers into the Rolls-Royce fold in the 1960s and, while the state enterprise has since been joined by other significant entries into the aeroengine field, France did essentially the same in 1945 through the creation of SNECMA, an agglomeration of all former private-sector engine companies under the state banner. Echoing the MBT experience, the NICs entered limited aeroengine production largely under the auspices of the state and its representative production entity, a 'flagship' engine enterprise. Table 3.8 lists the principal turbine (embracing turbojet, turbofan, turboprop and turboshaft technology) engine makers in the world outside the Soviet bloc. The NIC entries are state MIEs--in the case of China they are actually integrated airframe and engine factory units--while those associated with the AICs are a mixture of

Table 3.8 : Turbine Aeroengine Manufacturers

Country and Firm: AIC	NIC
Australia: CAC	China: Chengdu
Belgium: FN	Harbin
Canada: P & WC	Shanghai
France: Microturbo	Shenyang
SNECMA	Xian
Turboméca	Egypt: AOI Engine Factory
West Germany: MTU	India: HAL
Italy: Alfa Romeo	Israel: Bet-Shemesh
Fiat	IAI
Piaggio	Romania: Turbomecanica
Japan: IHI	South Africa: Atlas
KHI	Taiwan: AIDC
MHI	Yugoslavia: Orao
Sweden: Volvo Flygmotor	
UK: Rolls-Royce	
USA: Allison	
Avco Lycoming	
Garrett	
GE	
Pratt & Whitney	
Teledyne CAE	
Williams	

Source: Derived from Jane's All the World's Aircraft 1984-85, (Jane's
         Publishing, London, 1984).

state and private enterprises. Outside of the USA,
no AIC is currently capable of maintaining more than
three enterprises in the turbine aeroengine field.
   Nowhere is the concentration trend more evident
than in the production of military aircraft. In
particular, the number of MIEs participating in the
design and manufacture of front-line combat aircraft
has been severely curtailed in recent decades. Table
3.9 scans the US situation since 1960. The categories of 'fighter and attack' and 'bomber and ASW'

Table 3.9 : Rationalisation of US Aircraft MIEs

Aircraft type	Companies:1960	1966	1976	1986
Fighter and attack	Douglas GD Grumman Lockheed LTV McDonnell North American[1] Northrop Republic[2]	Cavalier Cessna Douglas GD Grumman Lockheed LTV McDonnell North American Northrop	Cessna Fairchild GD Grumman LTV MD[4] Northrop[5]	GD Grumman MD Northrop[5]
Bombers and ASW	Boeing Douglas GD Lockheed North American	GD Grumman	Grumman Lockheed Rockwell	Grumman Lockheed Northrop Rockwell
Military transports	Boeing Cessna Douglas Fairchild GD Grumman Helio Lockheed	Beech Boeing Douglas Grumman Helio Lockheed	Boeing Fairchild Lockheed MD	Beech Boeing Gates Learjet Grumman Gulfstream Lockheed MD
AWACS or AEW	Lockheed	Boeing Grumman	Boeing Grumman	Boeing Grumman
Helicopters	Bell Boeing Hiller[3] Hughes Kaman Sikorsky	Bell Boeing Fairchild Hughes Kaman Sikorsky	Bell Boeing Hughes[6] Kaman Sikorsky	Bell Boeing Kaman MD Sikorsky
Trainers	Cessna North American Northrop	Beech Cessna North American Northrop	Beech Boeing Cessna	Beech Cessna Fairchild Piper[5]

Notes: 1. Absorbed by Rockwell in 1967
2. Absorbed by Fairchild in 1965
3. Absorbed by Fairchild in 1964
4. Combination of McDonnell and Douglas after 1967
5. Export only
6. Absorbed by MD in 1984

epitomise the sophisticated combat   end of the military aircraft spectrum. Beginning   with the former,
the demands of the Vietnam War (as inferred from the
1966 listing) actually encouraged further entries:
Cavalier and Cessna specifically geared themselves
to meeting the need for close-support strike

aircraft thrown up by that theatre of operations. However, the aggregate increase in the number of MIEs engaged in this activity between 1960 and 1966 masks the fact that a fighter specialist, Republic, had withdrawn from the business by the later year. The cutbacks following the war are evident in a 1976 listing which is devoid of Cavalier, Lockheed and North American Aviation. Moreover, by that time, Douglas was merely a minor participant in the field under the aegis of the MD corporate banner while, interestingly, Republic had emerged from the shadows in the guise of a Fairchild subsidiary. A decade more and the number of fighter and attack aircraft prime contractors had suffered further retrenchment, standing at five after the loss of Cessna, Fairchild and LTV and the reinstatement of Lockheed (courtesy of its 'unofficial' F-19 'stealth' programme). The declaration of winners of the Advanced Tactical Fighter prototype competition in 1986 served to confirm the primacy of Lockheed and Northrop as the leading USAF fighter specialists into the next century. The other main combat category, bombers and ASW aircraft, has had its participants reduced from five in 1960 to four in 1986. It is worth noting, however, that the 1980s witnessed the introduction of Northrop into the bomber field through its expertise in the 'stealth' technology required for the Advanced Technology Bomber (in fact, the firm had been heavily involved in experimental bomber work in the late-1940s). The non-combat military aircraft categories mostly overlap with civil aircraft classes and thereby allow firms to produce for both markets. For example, the relatively large number of entries in the military transport and helicopter categories is a reflection of the ability of firms to cross-subsidise their civil and defence programmes. All of the military transport manufacturers with the exception of Grumman and Lockheed have a much greater stake in commercial, rather than military, aircraft production. By the same token, the helicopter makers (excepting Kaman) and the trainer producers are also thoroughly committed to civil aircraft of the general aviation kind. Only in the area of AWACS and ECM has entry of firms into the market increased from 1960 and this fact is symptomatic of the huge increase in the role of defence electronics in weapons systems.

Of course, other countries must make do with aircraft programmes that are only a fraction of the size of those in America and the outcome in them is extreme concentration in combat aircraft production. As with MBTs, each country attempts to sustain a

Table 3.10 : AIC and NIC Fighter Aircraft Producers

AIC: Company[1]	Type	NIC: Company[1]	Type
France: Dassault	Mirage	China: state factories	J-6/Q-5
	Super Etendard		J-7
	Jaguar		J-8
	Rafale	India: HAL	Jaguar
			MiG-27M
W. Germany: MBB	Tornado	Brazil: Embraer	AM-X
	Eurofighter	Romania: CNIAR	Orao
Italy: Aeritalia/		Yugoslavia: Soko	Orao
Aermacchi	AM-X	Israel: IAI	Lavi
Aeritalia	Eurofighter		Kfir
Japan: MHI	F-15J	S. Korea: KAL	F-5E/F
Netherlands: Fokker	F-16	Taiwan: AIDC	F-5E/F
Belgium: SABCA	F-16	Turkey: TUSAS	F-16
Sweden: Saab-Scania	Gripen		
	Viggen		
Switzerland: F+W	F-5E/F		
UK: BAe	Eurofighter		
	Tornado		
	Jaguar		
	Harrier		
USA: GD	F-16		
Grumman	F-14		
Lockheed	ATF		
	F-19		
MD	F-15		
	F/A-18		
	Harrier		
Northrop	ATF		
	F-5E/F		
	F-20		

Note: 1. Only includes companies undertaking construction of all or most of the fighter's airframe and its final assembly.

single producer (though each producer need not be confined to a single project at any given time). To put the picture into perspective, Table 3.10 intimates that all of the non-American AIC fighter firms sum to a total of ten--just twice as many as there are undertaking the business in the USA alone--and many of these owe their survival to the co-production of US designs (e.g. MHI, Fokker, SABCA and F + W). Their dependence is often reflected in their shareholding structure with, for example, Northrop owning 20 per cent of Fokker and Dassault-Breguet maintaining a minority stake in SABCA. The NIC fighter producers are overwhelmingly dependent on licence manufacturing of AIC or Soviet designs and, as with defence industry in general, rely for the

most part on state enterprises. Each NIC sets up a
monopoly supplier which is confined, by and large,
to producing for the domestic market (note, however,
that China's state aircraft factories and Israel's
IAI have definitely broken out of this strai-
ghtjacket to enter export markets in a big way).
Clearly, then, concentration is the norm for the
prime contract side of defence production and this
is the case irrespective of whether one is considering the AICs or the NICs and regardless of the particular branch of defence manufacture under scrutiny. Yet, just as industrial concentration is
prominent in defence production, so too is the
obsession with technological advancement. Indeed, as
we shall see below, the two are interactive.

Technological Underpinnings of MIE Operations

In one major respect at least, MIEs and civilian
firms share a common experience; namely, they both
succumb to the dictates of life-cycle theory. One
leading defence writer lends credence to this assertion by claiming that there is now 'good reason to
suppose that the process of technological diffusion,
the so-called product life cycle, will characterise
the military as well as the civilian sector in the
future'.[57] Other authoritative students hit on a
major facet of life-cycle theory, its emphasis on
the crystallisation of a global division of labour,
and suggest that glimmerings of it are now discernible among the various actors undertaking defence
production.[58]

> What seems to be forming in world defense pro-
> duction is an international division of labor
> in which the technologically advanced countries
> favor production of sophisticated systems on a
> high value per unit and therefore still remun-
> erative basis, while the developing countries
> emphasise indigenous or licensed manufacture of
> older generation military systems that can be
> produced in greater numbers at a much lower
> cost per unit. Advanced and vintage systems
> satisfy different market segments, rather than
> competing directly.

What, then, is the mechanism that drives life-
cycle theory? In a word, it is the transition from
emphasis on R & D to an obsession with productivity.
The former affords a firm market advantages through
its ability to generate new products as a direct

spin-off from the activity of experimentation. Consider: initial advantages in product development ensures the firm a segment of the market which perhaps can be built upon provided that learning-curve effects and economies of scale are subsequently marshalled. When the latter prevail, the firm has switched from the state of relying on product development for its competitive advantages to becoming enmeshed in a situation where it is reliant on productivity advantages over its competitors. According to life-cycle theory, there is a definite sequence of events underlying the transition; a sequence full of meaning for industrial organisations and host communities alike. In other words, both the firm and the location of its productive activities change as the product 'life cycle' works itself out. Essentially, the theory has an aura of determinism about it: not just in terms of stipulating the conditions conducive to the innovation of products, but also in relating the corresponding factors leading to the 'birth' and evolution of firms. Put briefly, the theory assigns a life cycle to each industrial product which countenances its beginnings as an innovation, its development and acceptance as a marketable item and its eclipse, in due course, by more timely substitutes. All products evolve from an initial stage of experimentation in which product specifications are 'fluid' and the markets for which the item is intended are elusive and ill-defined. Matching product 'fluidity' is a process technology which is likely to be labour-intensive, 'brain'-intensive and replete with customised items flowing off the production line in single orders or, at best, small batches. In the fullness of time, product specifications are unanimously adopted and standard tooling becomes feasible. The outcome is a process wherein the early changeable product lines are replaced by standard products which change, if at all, only in minor incremental ways and which, thereby, liberate the firm from design and product innovation and enable it to focus on 'the progressively intense quest for efficiencies in production that do not threaten to make obsolete ever-higher investments in a given product design or its associated capital equipment'.[59]

Put otherwise, productivity gain has replaced R & D as the prime mover in the evolution of the firm as an industrial organisation, and process standardisation is the usual means by which this evolution is deemed capable of achievement. Yet, by dint of concentration on process technology rather than product technology, the firm runs the risk of becoming wedded to product lines which are experiencing increasing 'maturity' and which are bound, eventually, to expire when overtaken by 'senescence'. Failure to recognise the need for product innovation may, in the long run, lead to the firm's demise since the managerial attitudes that were once receptive to new ideas and experimentation are now focused--at the exclusion of everything else--on the production task at hand. Thus, the evolution of the firm as an 'organic' entity derives from its transformation from a pioneering, 'open' organisation to a mature, 'closed' one. In the beginning, new firms are perforce small owing to their dependence on the introduction and subsequent acceptance of a tentative product innovation. Equally, they are loose-structured and non-hierarchical in managerial composition for the simple reason that this form of organisation is the most fitting for radical experimentation and the ready grasping of the significance of innovations. The small, informal and highly experimental (and successful) aircraft design team of the Dassault-Breguet MIE is frequently cited as a case in point. Over time, however, market consolidation and increasing size swing a firm into the productivity-orientation phase and effect a change in the managerial structure which typically re-emerges as large, rigid and ultra-hierarchical. This is borne out by some occurrences in defence industry. In the aeroengine sphere, for instance, both Westinghouse and GE were held up as examples of the mature, hierarchical industrial organisation: reputedly, they embarked upon the development of aircraft turbine engines in a manner that presupposed the innovation could be regarded as a variant of the mature steam-turbine technology. It was only in the mid-1950s, after competition from the loosely-structured, product-oriented Rolls-Royce and Pratt & Whitney firms had compelled serious managerial reassessment, that GE was able to assert itself as a major player in the modern aeroengine industry. Westinghouse's inability to undergo comparable restructuring led to its abandonment of the aeroengine market at about the same time.[60]

As codified, life-cycle theory is divisible into six stages.[61] The first, 'conception', embraces the invention of a potential product. It is followed by the 'birth' stage which takes in the conversion of the idea of a potential product into a physical or concrete prototype. The third stage, that of 'childhood', accommodates the initial manifestation of an industrial organisation. The emergent firm is located near the potential market for

the new product and engages in devising a diversity
of product specifications appropriate for that market. Its process technology rests on the application
of general-purpose machines by highly-skilled workers and its goal is to have the new product accepted
in a market niche. 'Adolescence' succeeds childhood
and is characterised by a reduction in the number of
preferred product designs, the beginnings of spe-
cially-adapted machines for processing, and a desire
for expansion on the part of the firm expressed
through its willingness to adopt strategies aimed at
maximising market share. By 'maturity', the firm is
using large-scale production processes to turn out
large volumes of standardised products. Automation
of production lines is accompanied by a degrading in
the skill-mix of the workforce, and together, these
factors allow the firm to transfer its main factory
units to locations where productivity can be
enhanced through the minimisation of factor costs
(in practice, this often entails choosing 'offshore'
cheap-labour sites). Finally, the stage of 'senescence' is broached when the product is so standardised that not even incremental innovation is war-
ranted and, to cap it all, is so readily available
from a host of imitative producers as a result of
the diffusion of its standard process technology,
that the original firm is obliged to divest itself
from undertaking the product's manufacture in order
to maintain profitability. At that juncture, the
firm either fades from the scene or, stimulated by a
Schumpeterian environment of 'creative destruction',
benefits from a new lease of life by virtue of its
ability to divert profits into a revived R & D
effort and, hence, new product innovation (as we
shall see, the Singer firm is a salient example of
such a 'born again' happenstance). Provided the second and more optimistic scenario comes to pass, a
truly global division of labour is in the offing.
The original market sites--overwhelmingly located in
the AICs--host all developmental stages where R & D
is in vogue, whereas the locations enjoying factor
advantages (especially comparative advantages in
labour costs) that become increasingly important as
production is standardised, tend to acquire industries occupying the mature and later stages of the
product life-cycle. For obvious reasons, then, the
NICs come to garner the lion's share of the current
standard product and process technology industries
while the AICs preserve whatever are the contemporary technology-intensive (innovation prone) industries for themselves.
　　The telling phrase is the asseveration that the

AICs will retain their competitive trade, industry
and corporate dominance by dint of specialisation in
technology-intensive activities. Such technology
intensity, of course, comes to the fore in defence
production and this fact alone accounts for the
tardy record of defence industry when it comes to
the transfer of manufacturing activities to NICs.
In marked contrast to much secondary industry,
defence technology is not succumbing to 'maturity'
in the sense that products can no longer gain from
technological (innovation) embodiment. If anything,
the requirement for R & D is accelerating. 'Not only
is the general rate of change increasing, but it
changes faster for aerospace technology and computers than for rifles and tanks' and it is in the
defence fields of aerospace and electronics that the
AICs doggedly hold on to their manufacturing lead.[62]
Thus, the more sophisticated combat aircraft require
replacement every ten to fifteen years while the
less sophisticated warship hulls and armoured fighting vehicles can stand service of twenty years or
more: the duration of the replacement cycle, therefore, is merely symptomatic of technology intensity
with the more complex systems requiring earlier
replacement.[63] And what is more, the cycle is not
only shorter for the more complex weapons but it is
becoming foreshortened in an intergenerational manner as well. Complexity is neatly encapsulated in
the proxy of R & D expenditures devoted to successive weapons systems for each defence sphere: in
sum, the replacement system incorporates proportionately more R & D than its immediate predecessor. The
progression of US bombers developed from the end of
World War 2 until the early-1960s unequivocally
bears out this proposition. The Convair (GD) B-36 of
the immediate postwar years required R & D expenditures amounting to two per cent of total weapon system life-cycle costs, the Boeing B-47 which followed
in the early-1950s had R & D bills equivalent to
three per cent of total costs; its immediate succes-
sor--the Boeing B-52--saw such costs boosted to a
figure of five per cent, while the aborted follow-on
project represented by the North American Aviation
(RI) B-70 was estimated to require R & D costs in
the order of 30 per cent of presumed life-cycle
costs.[64]
    Increasing technological embodiment goes hand
in hand with increasing unit cost of a weapon system
type. Nowhere is that trend more evident than in the
succession of US fighter aircraft subjected to
series production runs since World War 2. The flya-
way cost per production aircraft for the Republic

F-84D/E built in the years 1948-51 was $212,241. The firm's replacement F-84G of 1951-3 cost $237,247 while its swept-wing incremental improvement, the F-84F, was costed at $769,330 for a production run which extended into 1957. Republic's new-generation supersonic F-105, built after the F-84 model, jumped to a cost of $2.14 million in its D-marque manifestation. As for North American Aviation, its F-86A of 1948-51 cost $178,408 while the F-86F of 1952-5 weighed in at $211,111 and the supersonic spin-off, the F-100C/D models of 1955-9, were priced at more than three times this figure. Lockheed's F-94B of 1950-2 vintage cost the USAF a sum of $196,248 per copy but its F-94C improvement required a vastly inflated sum of $534,073 each; and its supersonic follow-on, the F-104A, required a cool $1.7 million per aircraft. For its part, GD's F-102A cost $1.2 million in the 1955-8 period whereas its upgraded successor, the F-106A of 1956-60, cost no less than $4.7 million for each aeroplane.[65] These trends, compounded since the 1960s in harmony with Kaldor's 'baroque' tendencies, have served to underscore the need for both innovation and the retention of extensive R & D facilities in MIEs located in the AICs.

It is hardly surprising, therefore, that defence production should conform to a pattern of locational inertia (recall Table 3.2); which is to say, once the MIE has set up its production facilities it is loth to dismantle, replace or relocate them. In short, the production side of advanced weaponry is too integrated into design and development to justify division of the organisation for the sake of offshore decentralisation of the manufacturing functions alone. Only failure of the follow-on system to come into effect and the resultant jeopardising of the MIE's very future provides the firm with sufficient incentive to expunge facilities and, thereby, revoke locational inertia. And even in this extremity, collapse of the corporate entity may not lead to the dissolution of physical assets: indeed, a bail out and corporate restructuring may guarantee the continuance of factories without impairing actual activity levels. The liquidation of Britain's most famous airframe manufacturer, the Sopwith Aviation Company, in 1920 as a result of difficulties associated with demobilisational instability had little material impact on its Kingston factory which, phoenix-like, rose from the corporate ashes as the main plant of the new Hawker enterprise (and, incidentally, survives to this day as the BAe Harrier production facility). Similarly, the collapse of the multinational (albeit UK based) Fairey

group in 1977 had little long-term effect on its
Belgian affiliate, Fairey SA of Gosselies, which
simply continued to function as a part-state enterprise called SONACA. It is safe to declaim, then,
that so long as that portion of the defence market
germane to a toppling MIE is not too crowded, the
actual production units of the business organisation
are likely to survive corporate failure either
totally untouched (e.g. Rolls-Royce's transformation
into government-owned Rolls-Royce (1971) Limited in
a classic case of permanent receivership was not
accompanied by any cuts in aeroengine factories) or,
at worst, after an interim period of inactivity (as
with Republic's Farmingdale plant which was converted from a fighter facility to one making attack
aircraft after acquisition and revival by Fairchild).

The reason for this state of affairs is
scarcely obscure: as units of capital--both physical
and human--the plants are repositories of manufacturing and design expertise which are regarded as
vital 'strategic' assets by the state. The ownership
of the plants is purely incidental to this fact.
Moreover, the R & D and intellectual capital care-
fully built up by the MIE is extremely sensitive to
activity disruptions and is prone to the risk of
dispersal--and loss to the state--should the facilities suffer closure. The Canadian case is salutary
in this respect. Closure of the Avro Canada MIE as a
direct result of the government's 1959 decision to
cancel the CF-105 Arrow high-performance fighter not
only denied Canada its only indigenous combat aircraft design team and R & D facility, but abrogated
the fruits of technology spin-offs to the country's
aerospace industry for a considerable period afterwards. A twelve-year famine in warship ordering had
comparable effects on the Canadian shipbuilding
industry, driving many shipbuilders out of the naval
market altogether and compelling others to re-enter
it only after incurring painful adjustment costs.[66]
Reputedly, the US space programme and its participant firms, based in the main in California, were
the only (and wholly unintended) gainers from the
windfall afforded them by the flux of unemployed
designers and engineers. Alive to the dire consequences falling out from the Avro Canada case, most
contemporary polities ensure that their prime
defence R & D and production facilities remain in
being, regardless of the MIE structure to which they
are attached at any given time. States undertake
this function solely because they cannot easily
replace the high-technology capabilities embodied in

such assets. On first appearances, then, it would
appear that defence industry will not readily follow
civil industry in the rush to move to NICs: their
symbiotic connections with the AIC state constitute
a political stumbling block for any such action in
the first place and, secondly, their continual need
to innovate weapons systems according to the dictates of the technological imperative conspires to
guarantee most MIEs a bout of renewal should 'senescence' set in with respect to older-generation weapons production.

That said, however, it is patently obvious that
some defence products are to all intents and purposes well within the bounds of standardised production characteristic of the 'maturity' phase of the
product life-cycle theory. By an interesting twist,
though, these are rarely earmarked by an AIC-based
MIE to one of its subsidiaries for production in a
NIC prior to re-import into the country hosting the
bulk of the MIE: rather, the technology pertinent to
a standardised product is licensed to a NIC state
enterprise so that the item may be produced for
domestic consumption in the licensee's country.
Sometimes, although infrequently, the advanced-coun-
try MIE may go so far as to create a subsidiary company in a NIC for the purpose of serving local markets should the NIC prove amenable to such action.
An arrangement of this kind may be forthcoming if
the NIC recognises its own technological and organisational shortcomings and is agreeable to the import
of expertise furnished by the foreign MIE. Examples
of technology transfer through the process of
licensing by firms in AICs to their state counterparts in NICs are legion, and the aerospace industry
in particular is replete with cases of licence-production agreements between US, European and Soviet
MIEs on the one hand and NIC producers on the other.
By way of contrast, the contemporary establishment
of MIE subsidiaries overseas is mainly directed at
the area of weapons platforms, that is to say, it
focuses on the provision of relatively unsophisti-
cated defence technology well within the capability
of local NIC workforces. Thus, the production of
fast attack craft, the most rudimentary form of warship technology, became feasible in South Korea when
the US Tacoma Boatbuilding Company erected a shipyard at Masan for the Korea Tacoma Marine Industries
enterprise in the late-1960s. By the same token,
West Germany's Lürssen MIE produced the first of its
twelve PZ-class armed patrol boats in 1981 for the
Royal Malaysian Police from the Butterworth, Penang,
yard of Hong Leong-Lürssen. Singapore, likewise,

had such a facility in the shape of Vosper Thornycroft Private Ltd—a subsidiary of the UK's Vosper
enterprise—until the yard was forced into liquidation in 1986. Besides minor naval shipbuilding,
there are many historical examples of foreign ventures by MIEs in other branches of defence production, but these are reserved for later consideration. Genuine MIE involvement in NICs for the
purpose of 'export platform' production is confined,
by and large, to the making of subcomponents. While
incontrovertible evidence is difficult to assemble,
some students opine that the 'main (US) producers of
military electronics have fairly extensive international operations' and that this fact alone is tantamount to their acceding to the process of decentralisation overseas of subcomponent manufacture in
order to realise factor-cost advantages.[67]

A GENERAL SCHEMA FOR THE MIE

The push and pull between specialisation and diversification in product manufacture is a stark choice
confronting the MIE throughout its corporate existence. As the life-cycle theory postulates, specialisation comes to the fore in the transition towards
the mature organisation. It is reinforced by the
fact that 'intimate familiarity with an existing
technology creates a strong disposition to work
within that technology, and to make further modifications leading to improvement rather than its displacement'.[68] By virtue of the technology-intensive
nature of defence R & D, it is reasonable to suppose
that the MIE will be most reluctant to abandon a
line of costly product innovation when a market
niche is within reach. Moreover, governments may
wish to preclude dissipation of effort on the part
of the MIE in order to keep alive the specialist
technological expertise embodied within it. State
enterprises may thus receive official sanction as
specialist producers with a guaranteed market and,
at least in their NIC manifestation, can circumvent
the innovation obstacle through purchase of technology from elsewhere. Private MIEs—and those state
enterprises disabused of the cosy notion of permanent receivership—may not be so fortunate. The
prospect of failing to gain a production contract
for the MIE's brainchild in a competitive tendering
context obliges the firm to contemplate a diversified product base to fall back upon should such a
contingency arise. In any event, the persistent competition between military users of weapons systems

in the name of 'technological one-upmanship' means
that no fully-fledged MIE with design pretensions
can afford to rest on its laurels where innovation
is concerned. For MIEs anchored to locations in the
AICs, the maturity and subsequent stages of the
life-cycle process are likely to oversee a gradation
from extensive growth through intensification to a
phase of possible rationalisation. Extensive growth
is brought about by the addition of capacity (perhaps even new plant sites) to meet the needs of customers when demand for a product is in full spate.[69]
Declining demand requires adjustments--the so-called
intensification process--that are designed to bolster productivity without summoning corresponding
charges: labour lay-offs and incremental process
technology improvements could be an integral part of
this procedure. When the product borders on obsolescence, the capacity devoted to its production is
liable to be axed under the rubric of rationalisation. Whether or not rationalisation entails actual
plant closures depends on the firm's ability to
replace the product with alternatives; be they follow-on items aimed at the same markets or markedly
different products geared to other markets. Obviously, a diversified enterprise has a greater prob-
ability of accomplishing a smooth transition from
one product line to another--and therein being able
to retain its plant capacity--than a firm committed
to a specialist product line. Consequently, MIEs are
mindful of the advantages of diversification and
have attempted, throughout their history, to broaden
the range of products they have to offer.
    One strategy is to opt for horizontal integra-
tion, that is, the acquisition through takeover or
merger of firms supplying the same market with comparable products. The opportunity to give full play
to economies of scale may underscore such moves
towards horizontal integration. On their own, individual producers may be incapable of realising
plants displaying the minimum efficient size (i.e.
the minimum size at which long-run average costs are
lowest) for the product line in question. The merging of firms and the deletion of small units in favour of the expansion of select large plants puts the
minimum efficient size criterion within reach of the
new, horizontally-integrated, MIE. It has been estimated, for example, that a plant just half the minimum efficient size for aircraft production in the UK
would incur cost overheads 20 per cent or more
greater than its twice-as-large, optimum counter-
part. As only one optimally-sized plant is needed to
fulfil the UK's demand for any particular aircraft

type, it made sense in the 1970s to merge both of the combat aircraft MIEs (Hawker Siddeley and the British Aircraft Corporation) into one organisation (BAe), concentrate production as far as possible at a select site, and divest the others of resources.[70] Horizontal integration, it should be noted, can also be undertaken for defensive reasons, that is to say, so as to co-ordinate erstwhile competitors in the business of dealing with a monopsonistic customer. Co-ordination in product development, manufacture and marketing, it is affirmed, will work towards providing the new enlarged MIE with improved (monopoly) prices, slash overheads with the demise of wasteful competition, and allow for the deletion of redundant capacity. The 1927 merger of Vickers and Armstrong Whitworth which gave birth to the UK's dominant warship and ordnance combine remains a classic instance of this option. Yet, horizontal integration may not be able to adequately cope with the problems of technological change. In short, its principal flaw is that it affords no escape route for a firm that remains a specialist—albeit a big and perhaps monopolistic supplier—when the technology in which it specialises becomes out-of-date or otherwise unwanted. That fate overcame the Vickers-Armstrong combine in its first decade: government reluctance to order capital ships so undercut demand for the MIE's hull-constructing, gun-making and armour plate manufacturing facilities that it was scarcely able to weather the depression despite harsh and deep cuts in capacity.

The alternative of vertical integration offers the prospect of common ownership of all the stages involved in the sequence of production right from the raw material origins of a product through to its disposal as a producer or consumption good. In undertaking vertical integration, a firm gains superior control over the scale and timing of production and thereby can realise considerable cost savings. Adolescent firms find it appealing since it allows them to integrate downstream (acquire forward linked activities) and so exploit learning-curve economies to attain their goals of dominating certain markets. Electronics firms of the likes of TI, Hitachi and NEC have been accused of following such a model.[71] Mature firms, meanwhile, are eligible to pursue vertical integration as a defensive ploy. Stable markets offer them the chance of profiting from productivity gains and then using their commanding position in the market to amass suppliers strung out along the production chain, pre-empting potential or real competitors that might be tempted

to do the same. An astute firm, however, will want
to disengage from vertical integration when the market approaches saturation point: if it does not,
exit barriers can lock that firm into a slow-growth
or declining market where it is duty-bound by signed
and binding contracts to support a veritable super-
structure of linked activities, all ailing in uni-
son. Its ability to shift into newer and more lucrative product lines is severely circumscribed as a
result. The fact remains, though, that vertical
integration is restricting in the same sense as horizontal integration; namely, it tends to confirm the
specialised nature of the firm with all that the
specialisation connotes in terms of exposure to
risks. It is worth bearing in mind that both Vickers
and Armstrong Whitworth had developed strong vertically-integrated organisations prior to their decision to amalgamate, but vertical integration had
been no more capable of saving them from the risks
of vanishing market demand than what was to tran-
spire under the subsequent horizontal integration
strategy.

The only reasonably safe means of overcoming
the risks intrinsic to specialisation is to diversify into increasingly-unrelated activities. The
corollary of this action is the important side-effect of size begetting size. In other words, unrelated diversification results in the firm becoming
larger and better able to direct resources to specific markets. Size and industrial concentration
that so ensue have distinct production, marketing
and risk-reduction advantages. Diversification of
this kind may occur as a grand horizontal integration exercise taking in all facets of defence production or it may work towards a conglomerate organisation which indulges in defence industry but much
else besides. The quintessential example of the
former is GD with its important stake in all three
of the principal defence markets: advanced combat
aircraft, major warships (nuclear submarines) and
MBTs. Examples of the latter are more plentiful, a
sure reflection of the greater risk reduction commensurate with a foot in the civil as well as the
defence camp. A seminal defence firm, Vickers PLC,
is now heavily engaged in a number of civil markets.
Since the shipbuilding and airframe interests were
shorn from it in 1977, the centrepiece of the firm's
defence activities has been military vehicles work;
that is, making its own series of MBTs for foreign
powers and Challenger armoured repair and recovery
vehicles for the MoD, as well as manufacturing turrets for the Warrior infantry combat vehicle. Other

defence business resides in the activities of the
marine engineering section (including orders for
stabilizers from the US Navy). In all, though,
defence and aerospace amounted to a relatively small
part of Vickers' current operations: in 1985 it provided sales of £69.5 million in comparison with the
£175.8 million of motor cars (i.e. Rolls-Royce
Motors), the £118.2 million of lithographic printing
plates and supplies, the £84.7 million of business
equipment and the £70.6 million of marine engineering. Sales of a lesser significance than defence
stemmed from medical and scientific equipment (£34.4
million) and the 'others' category (£49.1 million)
which, at the time, embraced printing and packaging
machinery, machine tools and special steels.[72] Significantly, the sales of defence and aerospace products (and the profit before interest too) amounted
to just twelve per cent of the group total, a
sharply reduced figure from that pertaining to most
of the firm's history. But still, the balancing of
defence and non-defence interests preoccupies conglomerate firms. In 1985-6, for example, Textron
added the Avco aerospace company to its other
defence/aerospace interests (exemplified by Bell-
Helicopter-Textron) for $1.4 billion and supple-
mented it with the purchase of Ex-Cell-O Corporation
for $1 billion, a firm that not only served the
aerospace industry with components for aeroengines
but also was an important supplier to the automotive
industry.[73] Not least of its assets in the latter
area is the Cadillac Gage Company of Warren, Michigan: a major manufacturer of armoured cars and AFV
gun turrets.
   The interplay between aerospace and motor vehicles is hardly coincidental: the first is geared
for the most part to defence markets leaving the
second to offer the useful counter-cyclical function
of reacting to civilian business cycles. Thus, the
classic 1967 merger of North American Aviation from
the aerospace field and Rockwell-Standard from the
automotive field was mediated because each firm
wished to benefit from the counter-cyclical properties afforded by the other. Yet, as if to emphasise
the persistent need for breadth in diversification,
that newly-merged conglomerate embarked, from the
next year, on a programme of acquisition.[74] The Tex-
tile Machine Works of Reading, Pennsylvania, the
Acme Chain Corporation of Holyoke, Massachusetts,
and the Hatteras Yacht Company (builder of luxury
cabin cruisers); not to mention 'mainstream' aerospace (Remmert-Werner) and automotive (the Luber-
Finer diesel engine-filter maker) companies all

succumbed to Rockwell bids in 1968. In the following year, Rockwell bought Miehle-Goss Dexter (manufacturer of printing and graphic arts equipment), Morse Controls (cable control systems) and Whittaker Marine (builder of fibreglass houseboats). By 1972, it had landed Collins Radio, a major civil and defence avionics manufacturer. At that juncture, automotive components accounted for 27.8 per cent of RI's total (1973) sales, aerospace contributed 32 per cent of the sales, the utility and industrial division added 22.7 per cent and electronics a further 14.1 per cent, leaving a residual of 3.4 per cent attributable to power tools.[75] Its subsequent moves into consumer electronics (through Admiral TV) proved disastrous and, in spite of its undoubtedly diverse product base, RI continues to rely inordinately on defence markets. Other automotive industry firms--GM, Ford, Chrysler, Fiat and Daimler-Benz, to name just a few--have also formed symbiotic relationships with defence industry over the years, as we shall see in due course.

Some MIEs aspire to achieve the counter-cyclical property using different civil market alternatives. At a simple and related diversification level, warship builders can convert to merchant shipbuilding and ship-repair. At the conglomerate level, the choices are far wider. One of the most prominent defence prime contractors of the 1980s, UTC, is fundamentally a conglomerate of major proportions. Its defence interests are most obvious through the activities of Pratt & Whitney and Sikorsky. The former is vital to the world's air arms as a supplier of aeroengines whereas the latter is one of the clutch of firms dominating the global military helicopter market. Both subsidiaries (and their subsidiaries such as P & WC and Westland) offer counter-cyclical work in civil aeroengine and helicopter markets. From 1975, however, and the acquisition of Otis Elevator, the UTC group has been adamant in its desire to diversify into unrelated areas. Three years later, the Carrier Corporation was brought into the fold and UTC plunged into the air-conditioning business. In the 1980s the manufacture of lifts was enhanced through the merger (in 1984) of the firm's Italian subsidiary, Stigler Otis, with Falconi in the first place and the formation of a joint venture with the Tianjin Elevator Company (in 1985) to undertake lift and escalator production in China in the second. At the same time, UTC began the manufacture of air conditioners and refrigeration units in Taiwan through the Carrier Taiwan Company. Other joint ventures included one with Toshiba of Japan to develop and make small power-generating units and another with Grundig of West Germany and Renault of France for the manufacture of automobile parts. In the core area of aerospace, UTC began 1986 with the partial acquisition of the Westland helicopter concern in the UK, the establishment of Pratt & Whitney of China to maintain and, eventually build, turbine aeroengines in that country, and, in partnership with SAI, the purchase of the former TRW subsidiary--Turbine Overhaul Services of Singapore--a repair facility for aeroengines.[76]

Clearly, then, any MIE belongs to a particular organisational setting--be it specialist or diversified, large or small scale--and owes its viability to the level of technological know-how it has at its fingertips. Equally obviously, an MIE that has undertaken incursions into several areas of defence production has a varied stock of technological expertise. For example, Lockheed may be a dominant player in terms of military aircraft and missiles, but only a relatively minor player in warship design and construction. When considering technical capability and hence, survivability of any MIE, it is important to note that classifications, appraisals and prescriptions of enterprise performance apply only to specific defence product lines: the firm's resilience deriving from diversification strategies, especially into civil markets, cannot be adequately gauged from the record of defence production alone. That proviso also precludes serious assessment of the firm's role as a comprehensive defence producer since, in light of the implications of product life-cycle theory, a MIE's involvement in one particular defence product line may have worked its course whereas its involvement in another area of defence production (provided the firm is diversified) may yet have barely moved beyond innovation and design. As most MIEs assiduously clamour for diversification, the latter situation is likely to be the more apposite one for any given defence contractor. In any case, even specialist defence firms will display a predilection for innovating replacement products capable of insertion into existing production lines if only to stay the course imposed by the technological imperative of defence markets. Therefore, it is fair to assume that a typical MIE will belong, at one and the same time, to the mature or senescence stages of the cycle with reference to some products, as well as asserting its right to future defence markets through participation in the birth and adolescent stages associated with other products.

Figure 3.1 : MIE Technology Capability Spectrum

	LICENCE PRODUCTION	INCREMENTAL IMPROVEMENT ON LICENCE PRODUCTION	LIMITED INDIGENOUS DESIGN	FULLY-FLEDGED DESIGN
WARSHIPS	FFG 7: Bazan Williamstown DY	GODAVARI: Mazagon Doc	LUDA: CSSC	FFG 7: Bath IW LEANDER: UK yards KOTLIN: USSR yards
COMBAT AIRCRAFT	MiG-27: HAL F-16: TUSAS	KFIR: IAI CHEETAH: Atlas	J-8: Shenyang	MiG: USSR factories MIRAGE: Dassault F-16: GD
MISSILES	IMPROVED HAWK: Mitsubishi Electric	BRAZILIAN COBRA: Army	KUN WU: Chung Shan	SAGGER: USSR factories COBRA: MBB HAWK: Raytheon
TANKS	LEOPARD 2: Contraves	TYPE 69: Chinese factories	OSORIO: ENGESA	T-54: USSR factories LEOPARD: Krauss-Maffei

With that in mind, Figure 3.1 presents a schema for categorising MIEs according to technological capability. The idea is to assign the MIE to the capability class typical of its main product line. It almost goes without saying, however, that some firms will straddle classes should their defence involvement entail significant commitment to more than one major product line. This qualification will be invoked whenever appropriate. At the lower end of the capability spectrum is the category of licence production. While most characteristic of NIC entries into defence production, licence manufacture is also widely followed by well-established MIEs in AICs (and elsewhere) which, for a variety of reasons, choose not to embark on development of certain weapons systems. Figure 3.1 instances the two warship-building state enterprises of Spain's Bazán and Australia's Williamstown Dockyard, both of which are constructing US-designed FFG7-class frigates (a minimum of four in the first case and two in the second). For combat aircraft, the category is aptly represented by the construction of Soviet MiG-27Ms in India by state-owned HAL and the impending construction of GD F-16s in Turkey by the state-inspired TUSAS organisation. The abundance of missile licence production arrangements is encapsulated in the single case of Mitsubishi Electric's manufacture of the US Improved Hawk SAM. In reference to tanks, the Swiss firm of Contraves (an affiliate of Oerlikon-Bührle) is leading the group charged with the licence production of West German Leopard 2 MBTs for the Swiss Army. Moving up a notch, the category of incremental improvement on licence production takes in MIEs able and willing to adapt foreign designs to better fit indigenous requirements. India's Godavari-class frigates are symptomatic of this phenomenon in as much as they are extensively-modified variants of British Leander-class vessels. At least three of them are to be forthcoming from the Mazagon Dock government undertaking in Bombay. The Israeli Kfir, a product of state-owned IAI, and South Africa's Cheetah, resulting from the efforts of state-owned Atlas, are both major improvements on the original Dassault-Breguet Mirage III fighter. In the missile arena, the Brazilian Cobra is a spin-off of West Germany's Cobra anti-tank weapon underwritten by the Brazilian Army's Institute of Research and Development in Rio de Janeiro while, in the tank field, the Type 69 is a Chinese incremental improvement of the Type 59, itself a version of the Soviet T-54 MBT.

The limited indigenous design category covers MIEs capable of developing all but a modicum of the components necessary to bring a weapon system to fruition. China's Luda-class destroyers and the state MIE that created them is a case in point. Owing only a trace of inspiration to the Soviet Kotlins of the 1950s, the Ludas are substantially larger and about 17 have been constructed by CSSC at Luda (Dalian), Guangzhou and Shanghai.[77] The aircraft factory at Shenyang was also responsible for a design with convoluted Soviet ancestry, the J-8 fighter. A twin-engined upgrade of the J-7, itself an improved version of the MiG-21, the J-8 also incorporates MiG-23 technology. Across the Formosa Straits in Taiwan, the Chung Shan Institute of Science and Technology has produced the Kun Wu anti-tank missile, supposedly influenced in its design by the Soviet Sagger. As far as tanks are concerned, Brazil's ENGESA, a private company, has developed the Osorio MBT with the help of the UK's Vickers (for the turret), Dunlop Aviation (for the hydro-pneumatic suspension system) and RO (for the gun).[78] The most capable MIE category, that of fully-fledged design, is indicative of the all-round contractor, equally conversant with innovation, development, manufacturing and overhaul. The MIEs belonging to this category that are mentioned in the figure are there because they have some bearing on the MIEs thrown up by the other categories. To this end, the US lead yard for the FFG7-class--Bath Iron Works of Bath, Maine--is present (having built or building, along with Todd Shipyards of San Pedro, California, no fewer than 51 for the US Navy). Moreover, both Dassault-Breguet and GD are present in their capacity as leaders in global fighter design, whereas MBB and Raytheon represent top-line missile MIEs and Krauss-Maffei (now a MBB subsidiary) enters the lists as one of the world's leading exponents of tank design.

Shifting to the organisational schema for MIEs, Table 3.11 concentrates on representative participants in the prime contract field of aircraft 'weapons platforms' (i.e. aircraft exclusive of their systems). The MIEs are divided in order of sophistication, with the most sophisticated 'lead prime' being offset by a number of dependent producers; namely, 'co-operative primes' (those responsible for part of the design but not its entirety), 'dependent co-producers' (those that modify existing designs or innovate modest design features) and 'licence producers' (those building-to-print without any design input at all). End-users are the military customers and are demarcated into single and multiple state instigators of weapons systems design and/or demand.

Table 3.11 : Organisational Schema I: Aircraft

End-user:	Lead prime	Co-operative primes	Producers: Dependent co-producer	Licence-producer
Single state hosting producer	MD (F-15)	Rockwell/LTV (B-1B)	IAI (Lavi)	MHI (F-15)
Multiple states hosting producer				
(a) state-induced	GD (F-16)	Aeritalia/Aermacchi/ EMBRAER (AM-X) BAe/Dassault (Jaguar)	ENAER (Pillán)	SABCA/Fokker (F-16) HAL (Jaguar)
(b) commercially- induced	Northrop (F-5/20)	Pilatus/BAe (PC-9)	AIDC (new fighter)	F+W (F-5) KAL (F-5)

Within the latter, international ventures are subdi-
vided into those stemming from government initiatives and those deriving from the inspiration of the
MIEs themselves. MD with its F-15 fighter is a lead
prime motivated by the demands of a single state,
the USA. Similar demands also obliged two MIEs--RI
and LTV--to combine resources to produce the B-1B
bomber. While RI is principal partner in the programme, LTV is geared to manufacturing much of the
aircraft's fuselage. The state of Israel stipulated
the requirements for the Lavi fighter of IAI, but
the parent MIE was beholden to Pratt & Whitney for
the engine and Grumman for much of the machine's
materials structure. MD's F-15 crops up again as an
example of a design subject to licence production;
in this case by MHI in Japan for the country's Air
Self-Defence Force. State-induced international ventures are ably illustrated by the record of the
F-16. Under the auspices of GD, four European
nations participate in its production and two of
their MIEs--SABCA in Belgium and Fokker in the Neth-
erlands--are tasked with assembling the fighter in
Europe. For their part, Aeritalia, Aermacchi and
EMBRAER engage in the joint design and manufacture
of the Italian-Brazilian AM-X close-support fighter
while BAe and Dassault-Breguet were jointly charged
with producing the SEPECAT Jaguar tactical fighter,
examples of which are emanating from the licensed-
production efforts of HAL in India. The sole example of a state-induced dependent co-producer, ENAER,
is interesting. This Chilean government-owned undertaking makes the Pillán military trainer, an aircraft designed in the main by the US Piper firm.
Copies of the aircraft are assembled by CASA in
Spain for service with that nation's Ejercito del
Aire in return for Chilean purchases of the Spanish
state enterprise's C-101 jet trainer. Commercially-
induced international venture are best represented
by that doyen of 'private venture' MIEs, Northrop.
Its F-5 (and follow-on F-20) was designed specifically for foreign sales and it continues to be built
under licence in Switzerland by F + W and in South
Korea by KAL. Northrop is actively assisting Taiwan's AIDC in the formulation of a new fighter
replacement for the F-5. It ought to be stated that
Taiwan was the original target for F-20 sales but,
owing to protests lodged with the USA by China, the
plan had to be abandoned. Lest it be thought that
such ventures are the preserve of Northrop, the
European firms of Pilatus (the aerospace subsidiary
of Switzerland's Oerlikon-Bührle MIE) and BAe have
an agreement to jointly complete and market the PC-9

military trainer—as brought home by the aircraft's inclusion in the 1985 Saudi package deal signed by BAe.

Turning to the warship side of 'weapons platforms', Table 3.12 apportions MIEs along comparable criteria. As alluded to elsewhere, Newport News remains the supreme example of a 'lead prime' in warship production through its role as monopoly supplier of the US Navy's giant Nimitz-class nuclear-powered aircraft carriers. Furthermore, it co-operates with GD's Electric Boat Division in the series production of Los Angeles-class nuclear fleet submarines, although Electric Boat is the 'lead' firm in design terms. Responsible for Spain's new aircraft carrier—the Principe de Asturias—Bazán counts as a dependent co-producer by virtue of its reliance on assistance provided by the US firms of Gibbs and Cox, Dixencast, Bath Iron Works and Sperry. At least this is a step up on its status in Figure 3.1 as a licence-producer of frigates and, as such, serves as a cautionary tale for assuming that MIEs invariably occupy a single functional niche or category. Williamstown Dockyard, however, retains in this schema the same status it shared with Bazán in the figure; namely, that of an indulger in licence-production on account of its complete absorption in FFG7 work. State-induced international ventures are typified by the standard frigate concept. The Netherlands and West Germany both chose to adopt a standard frigate hull that was based on the former's Kortenaer design, a vessel which was first built by the De Schelde shipyard. Taken up by West Germany's Bremer Vulkan shipyard (and others), the design re-emerged as the Type 122 or Bremen-class, suitably altered to meet Federal German Navy requirements (and, incidentally, fulfil regional bail-out needs for work-starved shipyards). Conveniently for the schema, the Greeks have stated their intention of building Kortenaer-class frigates under licence at the Hellenic Shipyards Company in Skaramanga.[79] Thus encouraged, the Dutch also co-operated in the emergence of a standard minehunter, the Eridan or Tripartite-class: a type building in some numbers in France (by DTCN, Lorient), Belgium (by Beliard) and the Netherlands (by Van der Giessen de Noord).

Sandock Austral (retitled Dorbyl Shipbuilders in 1987) of Durban, South Africa, fits the category of dependent co-producer for international state-induced schemes. Its Minister-class fast attack craft are modifications of the Saar 4 missile vessels built by the Israel Shipyards concern. Finally, commercially-induced international ventures are

Table 3.12 : Organisational Schema II: Ships

End-user:	Lead prime	Co-operative primes	Producers: Dependent co-producer	Licence-producer
Single state hosting producer	Newport News (Nimitz)	Newport News/GD (Los Angeles)	Bazán (Príncipe de Asturias)	Williamstown DY (FFG7)
Multiple states hosting producer				
(a) state-induced	De Schelde (Kortenaer)	Bremer Vulkan (Kortenaer/Type 122)	Sandock Austral (Minister)	Skaramanga (Kortenaer)
	Israel Shipyards (Saar 4)	DCAN/Van der Giessen de Noord/Bellard (Eridan)		
(b) commercially- induced	Thyssen Nordseewerke (Santa Cruz)	Korea-Tacoma (PSMM-5)	AFNE (MEKO 140)	Astilleros Domecq Garcia (Santa Cruz)
	Vosper Thornycroft (Niteroi)			Arsenal de Marinha do Rio de Janeiro (Niteroi)
	Tacoma Boat (PSMM-5)			China SB (PSMM-5)

commonplace in the warship market. The West German
firm of Thyssen Nordseewerke, for example, is
assisting the Argentine Astilleros Domecq Garcia
enterprise in the build up of a fleet of Santa Cruz-
class diesel-engined submarines. The Argentine
shipyard, while being a state venture, is rendered
feasible entirely on account of the technology
transferred from Thyssen and this dependence is
reflected in the 25 per cent shareholding reserved
for the German company. In a similar fashion, the
UK's Vosper Thornycroft has successfully aided Brazil's Arsenal de Marinha do Rio de Janeiro in the
completion of the Niteroi class of frigates. Not to
be outdone, the US Tacoma Boat enterprise has transferred its PSMM-5 fast attack craft technology to
Taiwan so that series production can be got under
way at the China Ship Building Corporation. The same
basic vessel has been produced by the US firm's
South Korean subsidiary for export to Indonesia and,
in this guise, Korea Tacoma qualifies as a co-operative prime. In conforming to the dependent co-producer role, Argentina's AFNE is constructing all six
of the MEKO 140-type frigates required by the Argentine Navy while relying, all the same, on Blohm +
Voss of West Germany for the ship's design.

Examples of MIEs subscribing to each category
can be replicated over and over again, not merely
for aircraft and ship producers but for the makers
of missiles, tanks and a host of systems. In the
tank field, for instance, GD is of overriding importance, acting as a 'lead prime' through its production of M1 and M1A1 vehicles. It also was the principal designer of the South Korean K-1 MBT, a tank
under construction at Hyundai's Changwon plant (making Hyundai, in consequence, an example of a 'dependent co-producer').[80] OTO Melara, for its part, mer-
its designation as a 'co-operative prime' by virtue
of its recasting of the Krauss-Maffei Leopard 1 MBT
into a specialised tank for the Italian Army. The
'licence producer' category, meanwhile, is filled by
India's Heavy Vehicles Factory which, from 1965,
built over 1,500 Vickers Mark 1 MBTs at the Avadi
plant near Madras. That factory is now turning to
licensed production of Soviet T72M1 tanks. Its sis-
ter facility, the Armoured Vehicle Factory at Medak
near Hyderabad, is currently pre-occupied with manufacturing under licence the Soviet BMP-2 infantry
combat vehicle.[81] However, it will not profit us to
continue providing an inventory of MIEs conforming
to the various typologies: the distinctive attributes of each functional type has been sufficiently
embellished in the foregoing account. Instead, other

enterprises will receive attention as and when they warrant consideration under the terms of the remaining chapters of this book. Suffice it to say that, as MIEs can demonstrably be placed in a functional schema, the onus of study must shift to address the question of how the structure of that schema has been formed. In other words, an evolutionary approach to the emergence of the MIE, and its subsequent development, constitutes the subject matter of the next chapter.

NOTES AND REFERENCES

1. J. F. Gorgol, The military-industrial firm: a practical theory and model, (Praeger, New York, 1972). Nevertheless, we shall draw heavily on the ideas put forward by Gorgol in formulating our MIE.
2. J. K. Galbraith, quoted in J. Tirman (ed.), The militarization of high technology, (Ballinger, Cambridge, Mass., 1984), p.11.
3. D. Gold, 'Conversion and industrial policy' in S. Gordon and D. McFadden (eds.), Economic conversion: revitalizing America's economy, (Ballinger, Cambridge, Mass., 1984), pp.191-203. Quote from p.197.
4. G. Kennedy, Defense economics, (Duckworth, London, 1983), p.164.
5. J. E. Ullmann, 'The Pentagon and the firm' in Tirman, The militarization of high technology, pp.105-22. Quote from p.110.
6. D. Holloway, The Soviet Union and the arms race, (Yale University Press, New Haven, 1983), p.119.
7. R. Huisken, 'Armaments and development' in H. Tuomi and R. Väyrynen (eds.), Militarization and arms production, (Croom Helm, London, 1983), pp.3-25. Quote from p.21.
8. The arguments pro and con domestic defence production apply equally to the AICs. See K. Hartley, NATO arms co-operation: a study in economics and politics, (George Allen & Unwin, London, 1983), pp.82-7.
9. T. W. Graham, 'India' in J. E. Katz (ed.), Arms production in developing countries, (D. C. Heath, Lexington, Mass., 1984), pp.157-91.
10. V. S. Ram, N. Sharma and K. K. P. Nair, Performance of public sector undertakings, (Economic and Scientific Research Foundation, New Delhi, 1976). For a contemporary summary of their standing, see Jane's Defence Weekly, 20 June 1987, p.1326.

11.   P. MacDougall, Royal Dockyards, (David &
Charles, Newton Abbot, 1982), pp.166-7.
12. A. W. Saville, 'The naval military-industrial complex, 1918-41' in B. F. Cooling (ed.), War,
business, and American society: historical perspectives on the military-industrial complex, (Kennikat,
Port Washington, NY, 1977), pp.105-117. Quote from
p.106.
13. J. F. Crowell, Government war contracts,
(Oxford University Press, New York, 1920), p.32.
14. J. Goodwin, Brotherhood of arms: General
Dynamics and the business of defending America,
(Times Books, New York, 1985), p.99.
15. J. P. Miller, Pricing of military procurements, (Yale University Press, New Haven, 1949),
pp.116-118.
16.   M. White, 'General Dynamics boss quits',
Manchester Guardian Weekly, 2 June 1985, p.9. Note,
North American Aviation is now part of Rockwell
International.
17.   K. Hartley, 'UK defence policy: seeking
better value for money' in RUSI and Brassey's
Defence Yearbook 1985, (Brassey's Defence Publishers, London, 1985), pp.105-119.
18. H. O. Stekler, The structure and performance of the aerospace industry, (University of California Press, Berkeley, 1965), p.191.
19.   Note, Aviation Week & Space Technology, 14
July 1986, p.127. Indeed, Raytheon's 'second-source'
business has ballooned: in 1986 it was designated as
the supplementary producer to GM Hughes for the
Phoenix AAM (see Flight International, 1 November
1986, p.10). It is also acting as a second source to
GD for the Standard 2 SAM.
20. As mentioned in Aviation Week & Space Technology, 9 June 1986, p.13.
21.   J. R. Gordon, 'NATO industrial preparedness' in L. D. Olvey, H. A. Leonard and B. E. Arling-
ghaus (eds.), Industrial capacity and defense planning, (D. C. Heath, Lexington, Mass., 1983),
pp.35-63.
22. See Flight International, 24 May 1986, p.8.
23. M. M. Postan, British war production,
(HMSO, London, 1952), pp.342-3.
24.   Holloway, The Soviet Union and the arms
race, p.122. The USSR also pushes export orders so
as to boost production runs. For example, from
1976-81 as much as 46 per cent of combat aircraft
and minor naval vessels produced in the USSR were
earmarked for export, no less than 53 per cent of
major warships were so destined and about 44 per
cent of tanks and self-propelled guns went as

exports.
25. J. H. Hoagland, 'The US and European aerospace industries and military exports to the less
developed countries' in U. Ra'anan, R. L. Pfaltzgraff and G. Kemp (eds.), Arms transfers to the
Third World: the military buildup in less developed
countries, (Westview, Boulder, 1978), pp.213-27.
Also A. J. Pierre, The global politics of arms
sales, (Princeton University Press, Princeton,
1982), pp.97-100.
26. For a discussion of these economies see C.
Groth, 'The economics of weapons coproduction' in M.
Edmonds (ed.), International arms procurement: new
directions, (Pergamon, New York, 1981), pp.71-83.
Groth goes on to show that scale and learning-curve
economies may be mutually exclusive or, in the ideal
situation, can work in tandem with each other.
27. K. Hartley, 'The political economy of NATO
defense procurement policies' in Edmonds, International arms procurement, pp.98-114.
28. As noted in Flight International, 1 November 1986, p.13.
29. The enforced rationalisation of UK airframe
companies into two groups--the British Aircraft Corporation and Hawker Siddeley--in 1960 is a case in
point. See D. Todd and J. Simpson, The world aircraft industry, (Croom Helm, London, 1986),
pp.156-7.
30. J. R. Kurth, 'Why we buy the weapons we
do', Foreign Policy, no. 11 (Summer 1973), pp.33-57.
31. J. Gansler, The defense industry, (MIT
Press, Cambridge, Mass., 1980), p.30.
32. As reported in Aviation Week & Space Technology, 12 May 1975, p.17.
33. Note Flight International, 21 June 1986,
p.14.
34. Note Flight International, 8 November 1986,
pp.22-3 and Aviation Week & Space Technology, 10
November 1986, p.24. As a point of fact, no less
than 1,048 F-5s were sold for $959 million to 15
countries between 1962 and 1971: some 58 per cent
of them being paid for by the US Government as military aid. See I. Dörfer, Arms deal: the selling of
the F-16, (Praeger, New York, 1983), p.63.
35. C. D. Bright, The jet makers: the aerospace
industry from 1945 to 1972, (Regents Press of Kansas, Lawrence, 1978), p.199.
36. Dörfer, Arms deal, pp.32-3 and Gansler, The
defense industry, p.204.
37. K. Hayward, Government and British civil
aerospace: a case study in post-war technology policy, (Manchester University Press, Manchester,

1983), p.20.

38. As stated by then Minister of Technology Tony Benn and reported in Flight International, 22 May 1975, p.812.

39. B. Walters, 'French naval technology', Navy International, vol. 91 (1986), pp.585-96. Quote from p.592.

40. J. Moore (ed.), Jane's fighting ships 1986-87, (Jane's Publishing, London, 1986), p.121.

41. Note Aviation Week & Space Technology, 16 June 1986, p.104 and Flight International, 11 October 1986, p.14.

42. See Business Week, 17 March 1986, p.87. In point of fact, Gates sold its 64.8 per cent share of Learjet for $62.7 million in September 1986 to New York investment firm Rosenthal & Associates. Refer to Interavia, November 1986, p.1259. Difficulties faced in consummating the sale led to a new deal which placed Learjet in the hands of its own management and a New York financial services firm. See Aviation Week & Space Technology, 15 December 1986, p.27 and 10 August 1987, p.32.

43. R. Schlaifer, Development of aircraft engines, (Graduate School of Business Administration, Harvard University, Boston, 1950), p.136.

44. As reported in Navy International, vol. 91 (1986), p.616.

45. They were added to the ten already assigned to Richards Ltd for construction at Lowestoft and Great Yarmouth.

46. M. Pearton, The knowledgeable state: diplomacy, war and technology since 1830, (Burnett Books, London, 1982), p.109.

47. Note comments in Fairplay issues of 1 January 1903 and 24 March 1904.

48. D. Todd, The world shipbuilding industry, (Croom Helm, London, 1985), pp.330-6.

49. Hayward, Government and British civil aerospace, p.213.

50. E. A. Kolodziej, 'Determinants of French arms sales: security implications' in P. McGowen and C. W. Kegley (eds.), Threats, weapons, and foreign policy, (Sage, Beverly Hills, 1980), pp.137-75. Quote from p.151.

51. J. Chillon, J-P. Dubois and J. Wegg, French postwar transport aircraft, (Air-Britain, Tonbridge, 1980), pp.35-6.

52. See Aviation Week & Space Technology, 15 December 1980, p.19.

53. R. E. Bolton, Defense purchases and regional growth, (The Brookings Institution, Washington, DC, 1966), pp.143-7 and C. H. Dillon,

'Government purchases and depressed areas' in H. A. Cameron and W. Henderson (eds.), Public finance-- selected readings, (Random House, New York, 1966), pp.97-105.

54. B. Bluestone, P. Jordan and M. Sullivan, Aircraft industry dynamics: an analysis of competition, capital, and labor, (Auburn House, Boston, Mass., 1981), pp.82-3.

55. M. Kaldor, The baroque arsenal, (Hill & Wang, New York, 1981), p.193.

56. Gansler, The defense industry, pp.42-3.

57. S. G. Neuman, 'Third World arms production and the global arms transfer system' in Katz (ed.), Arms production in developing countries, pp.15-37. Quote from p.27.

58. R. W. Jones and S. A. Hildreth, Modern weapons and Third World powers, (Westview, Boulder, 1984), p.64.

59. W. J. Abernethy, K. B. Clark and A. M. Kantrow, Industrial renaissance: producing a competitive future for America, (Basic Books, New York, 1983), p.20.

60. B. H. Klein, Dynamic economics, (Harvard University Press, Cambridge, Mass., 1977).

61. R. U. Ayres, The next industrial revolution: reviving industry through innovation, (Ballinger, Cambridge, Mass., 1984), p.84.

62. R. G. Head, 'The weapons acquisition process: alternative national strategies' in F. B. Horton, A. C. Rogerson and E. L. Warner (eds.), Comparative defense policy, (Johns Hopkins University Press, Baltimore, 1974), pp.412-25. Quote from p.413.

63. W. Bajusz, 'International arms procurement, multiple actions, multiple objectives' in Edmonds, International arms procurement, pp.188-216.

64. R. J. Art, The TFX decision: McNamara and the military, (Little, Brown & Co, Boston, 1968), p.88.

65. M. S. Knaack, Encyclopedia of US Air Force aircraft and missile systems, (Office of Air Force History, Washington, DC, vol. 1, 1978).

66. T. G. Lynch, 'DELEX: The Nipigon experience', Navy International, vol. 91 (1986), pp.301-303.

67. H. Tuomi and R. Väyrynen, Transnational corporations, armaments and development, (Gower, Aldershot, 1982), pp.140-2.

68. N. Rosenberg, Inside the black box: technology and economics, (Cambridge University Press, Cambridge, 1982), p.129.

69. Terms operationalised by D. Massey and R.

Meegan, The anatomy of job loss, (Methuen, London, 1982), pp.18-20.

70. M. A. Utton, The political economy of big business, (St Martin's Press, New York, 1982), p.40.

71. K. R. Harrigan, Strategies for vertical integration, (D. C. Heath, Lexington, Mass., 1983), pp.315-6.

72. According to a statement on 1985 results issued in The Economist, 1 March 1985, p.12.

73. As reported in Business Week, 1 September 1986, p.34.

74. E. P. Hoyt, The space dealers, (John Day Company, New York, 1971), p.141.

75. M. S. Salter and W. A. Weinhold, Diversification through acquisition: strategies for creating economic value, (Free Press, New York, 1979), pp.204-216.

76. See Business Week, 24 March 1986, pp.88-9 and Flight International, 22 November 1986, p.60.

77. B. Hahn, 'Chinese navy: first destroyer construction programme 1968-85', Navy International, vol. 91 (1986), pp.690-5.

78. C. F. Foss (ed.), Jane's armour and artillery 1985-86, (Jane's Publishing, London, 1985), pp.1-3. France's GIAT was offering an alternative gun for the vehicle.

79. An undertaking now extremely unlikely in view of Greece's budgetary difficulties. Cheaper foreign designs are under consideration, as noted in Navy International, vol. 92 (1987), p.224.

80. As described in Jane's Defence Weekly, 31 October 1987, p.994.

81. See report in International Defense Review, November 1987, p.1557.

Chapter Four

THE EVOLUTION OF THE TRADITIONAL MIE

The evolving organisation of the MIE is incumbent in
the first place on the broad sweep of technological
change which has spat out a series of weapons systems, each technologically more demanding than its
predecessor. Yet, the MIE is dependent in the second
place on its ability to marshal the technical inge-
nuity necessary to master new weaponry and thereby
enter into defence production. Put succinctly, the
MIE arose partly as an innovator of weapons systems
and partly as the most effective means of commanding
the resources needed to put those systems into production. Thus, the MIE was at one and the same time
the originator of technologically-endowed weaponry
and the dependant of its specific and demanding production and organisational requirements. In order
to stay the course, MIEs were much given to technical experimentation in what amounted to a cumulative
process whereby innovation in increasingly more complex and sophisticated weaponry was perpetuated. The
outcome was the compulsive drive associated with the
term 'technological imperative' which imbued industrial, military and government circles alike. The
combination of innovative prowess, command of
resources and extraordinary industrial organisation
first came to the fore in the latter half of the
19th century when the MIE, as we know it, was born.
This early or 'traditional' MIE, invested with the
wherewithal to produce complex weaponry in quantity,
was characterised by ventures spun off from the
steel and mechanical engineering industries.
Indeed, a preoccupation with armour, heavy ordnance
and shipbuilding is incident to the traditional MIE.
Coeval with the blossoming of the 'second industrial
revolution' based on the internal combustion engine
and the widespread adoption of electric power and
traction is the beginnings of a new type of MIE;
namely, that geared to the application of motor

vehicles and aircraft to the needs of armed forces. Much stimulated by the demands of World War 1, these 'modern' MIEs had partially eclipsed their traditional counterparts by the eve of World War 2. In that conflict, however, they received a tremendous fillip and emerged from it as the jewels in the defence industry crown. Nevertheless, that war had served to stimulate the innovation of devices which would increasingly come to challenge the supremacy of the modern MIEs in the subsequent postwar period. Centred on the use of electronics, a spate of new entries into the defence field were to capitalise on radar, sonar and the whole gamut of the seemingly baffling arena of ESM, ECM and ECCM. In some respects the 'emergent' MIEs merely complement their modern and traditional precursors: after all, they are largely involved in making systems for incorporation into weapons platforms that are still unequivocally ships, AFVs and aircraft. However, these systems are now so complex and expensive as to outweigh the costs--and significance--of their platforms. Moreover, the emergent MIEs have evolved an entirely new weapon system, the missile, which owes its genesis as a credible battlefield tool to the use of electronics for its guidance. As demonstrated in the Vietnam War, a series of Middle Eastern conflicts, and hostilities around the Falkland Islands, the serious and deadly business of electronic warfare (EW) in conjunction with missile armouries constitute the contemporary 'leading edge' of defence capabilities and the emergent MIEs that specialise in them occupy the ascendant position in contemporary defence industry.

Such an evolutionary schema is replete, of course, with simplification and generalisation. It does, none the less, suffice to make a distinction between longstanding MIEs and new-entry firms which represent either the fruits of defence-industrial innovation (at whatever stage, the 'traditional', 'modern' or 'emergent') or the diversification of civil enterprises into defence production on the hopes of realising greater-than-normal profits from the opportunities occasioned by access to defence markets (perhaps reflecting the counter-cyclical phenomenon). It also offers scope for a longitudinal examination of the fortunes of well-established MIEs. Thus, the behaviour of firms as reflected through their strategies of growth, acquisition and rationalisation can be assessed in light of the continual need confronting them to maintain technological and market competitiveness. Firms may oscillate between activity and passivity in innovation, with the former corresponding to entry into production and the accumulation of market share, while the latter requires defensive strategies of vertical or horizontal integration and, perhaps, retrenchment. In the final analysis, successful firms are those able to stay the course without excessive diminution or dismemberment; those, in other words, that have demonstrated their ability to conform to the rigours of the technological imperative and convince the state monopsonist of their continued merit as prime contractors. Mindful of this ultimate accolade, the purpose of this chapter is to outline the development of traditional MIEs, review their activities during the heyday of international rivalry based on battleship deployment, and consider their subsequent fate given the by-passing of such technology by new weapons systems. The chapter concludes with a commentary on the current lot of the survivors, either disposed to diversify into newer defence technology (or retreat from the defence field altogether) or to subsist as clients of the state--perhaps formalised through state ownership--within the bonds of permanent receivership and increasingly vulnerable to competition from new-entry firms in the NICs.[1]

HISTORICAL UNDERPINNINGS

The battleship 'Vengeance' of the Admiralty's 1897-8 programme was the first Royal Navy vessel to be 'built, engined, armoured, and supplied with her heavy gun-mountings by one firm'.[2] The firm in question was Vickers and it epitomises the state-of-the-art in defence production at the end of the last century, a state embracing the 'combination of heavy engineering with precision-skills of delicate toolmaking' that went hand in hand with investment in very specialised plant.[3] This plant was extremely expensive ('lumpy' in economic parlance) and fitted only to the needs of defence production. It could not, therefore, be readily turned to civil products during defence downturns and its capitalisation effectively denied entry into defence markets to all but a select few firms located in what are now termed the AICs. Vickers, along with Armstrong, Krupp, Bethlehem and one or two more, had achieved the stage of 'maturity' in the era immediately preceding World War 1, a stage characterised by volume production of standardised guns and armour and quantity production of not-so-standardised warships--the platforms which accommodated the guns and armour.

These firms were truly general defence contractors, engaging in all three endeavours on a massive scale. They, and they alone, had built up the expertise and plant to manufacture heavy ordnance, armour plate and capital ships. The pains taken to produce the first of these is aptly brought out in the thought that heavy artillery pieces 'were prepared by 90-foot lathes and rifled by boring machines set on a 108-foot bed and designed to give each cannon 9-12 months of precisely calibrated and uninterrupted machining'.[4] For its part, armour-plate manufacture rested on the provision of giant hydraulic presses. The one installed by Beardmore at Parkhead in Glasgow in 1898 weighed in at 12,000 tons capacity and, as such, was larger than the two giant Sheffield presses of John Brown (10,000 tons) and Vickers (8,000 tons) and the Manchester press of Armstrong Whitworth (10,000 tons). William Beardmore's brand-new warship construction works at Dalmuir on the Clyde was equally indicative of the resources put into this branch of defence production. Began in the early-1900s, the shipyard covered about 90 acres, contained six building berths equipped with an overhead gantry, had a fitting-out basin which was the largest in the world, and boiler and engine shops which, at almost 5.5 acres in area, were among the world's biggest.[5] Facilities of this stature were expensive of labour usage as well, and 'defence enclaves' were an inevitable consequence of their establishment. The MIEs came to act as key industries in geographically-concentrated industrial complexes where they 'tended to dominate groups of smaller companies in related trades and particular regions, like North-East England, the West of Scotland, Northern Ireland, Sheffield and Birmingham'.[6] It is not coincidental that the commencement of industrial decline after World War 1 in those same regions occurred in tandem with the process of stagnation and retrenchment experienced by their key host firms, the traditional MIEs.

FROM CONCEPTION TO MATURITY

Until the arrival of the traditional MIE, military products had tended to be manufactured by state arsenals and dockyards using process technology which was to all intents and purposes practically indistinguishable from civil standards. It was only during periods of unusually heavy demand--what we would term wartime equilibrium--that the services of civilian contractors were extensively utilised. For

134

one thing, it was believed in official quarters that private firms were not up to the exacting standards of the state MIEs and, in any event, would attempt to pass off skimpy work when outside of the full supervision of state overseers. So far as the Royal Navy was concerned, the government made the momentous decision during the 27-ship programme of 1690 to extend the sources of supply, previously confined to the Royal Dockyards and a few trusted private contractors located proximate to Whitehall, outwards to take in suppliers elsewhere in the country. Capacity constraints among the existing suppliers led directly to orders being placed in Bursledon, Southampton, Hull and Harwich: a step undertaken with some trepidation by the authorities as their previous experience of private contracting had been less than auspicious. For example, the flagship of 1637, the 'Sovereign of the Seas', had the dubious honour of being built by a private contractor, Phineas Pett, after severe cost overruns. In fact, the estimated building cost of £13,000 was exceeded to the tune of £45,500 and gave the naval procurement officials much sober food for thought.[7] A practice grew up during the 18th century whereby the Royal Dockyards—and some Thames private shipyards which could be closely supervised by the Navy—acted as 'lead' yards, introducing new classes of ship, but leaving the 'country' yards to top up the class numbers during episodes of emergency fleet expansion. Indeed, at the time of the Seven Years War (1756-63), warships emanating from the private shipbuilding industry exceeded those produced by the state for the first time.[8]

By the end of the century and the massive expansion triggered by the French wars, the Navy had formulated a species of division of labour whereby the largest warships would remain the preserve of the Royal Dockyards while private yards would be earmarked for vessels of inferior size and complexity. These latter vessels were all built to a standard price per class with variable allowances for Thames-built ships (more expensive) as opposed to 'country'-built ships (less expensive). The differentials were forthcoming in recognition of cheaper production costs—both in labour and in procuring timber materials—outside of London. In point of fact, a tendering system was not introduced until late in the 19th century. Table 4.1 provides some idea of the dispersion of naval contracts in the 70 years preceding the innovation of 'ironclad' warships in 1860.[9] With the exception of a solitary vessel emanating from Milford Haven, all of the

Navy's prime capital ships (1st and 2nd rates) derived from Royal Dockyards; although it must be said that use was made of government facilities at Kingston, Canada, and Bombay, India. It was only with second-class battleships (3rd rates) that private contractors came to the fore. In fact, fully 56 per cent of these vessels was produced by commercial shipbuilders: the Thames alone being responsible for 38 per cent. Royal Dockyards built the bulk of the relatively unimportant class of 4th rates, but production of frigates (5th and 6th rates) was not only shared much more evenly with private contractors, it was dispersed to a much wider set of geographical sites. Yards on the Tyne and Humber were contracted to supply tonnage, albeit in relatively insignificant volumes in comparison with the tonnage supplied by yards in the South of England. The most numerous of the principal warship classes, the sloop, was built in overseas private yards as well as around the coast of England and Wales. As with other classes, however, the vast weight of tonnage issued from shipyards located fairly close to the main Royal Dockyards in the South of England. This fact was a valid reflection of their status as dependent contractors, functioning very much in the manner of licence-producers of first-of-class ships originating in the state yards. By no stretch of the imagination, then, could the private yards be conceived of as being genuine MIEs possessing a dedicated warship design capability and a specific prowess in defence technology.

Similar conditions prevailed with other types of military hardware. Governments maintained arsenals for the manufacture of small arms and artillery pieces: in the British case the Tower of London was the traditional source of the former and the cannon factory at Woolwich of the latter. Again, it was only the exigent demands of war that compelled the farming out of weapons contracts to private firms, as the practice of wholesale dispensing of musket and rifle orders to small workshops in London and Birmingham during the Napoleonic Wars clearly attests. Yet, as befits its role as founder of the industrial system, the UK was to herald fundamental changes in the organisation of weapons supply at about the time of the Crimean War in mid-century. The change was to come about for two reasons. The first was somewhat ephemeral in that it was an attempt to make good past defects in military organisation in the face of open hostilities with a major power (Russia) and potential hostilities with others. The second, however, was far more profound

Table 4.1 : British Warship Contracts, 1790-1859

Ship type[1]	Distribution (number)	
1st rate (100-110 guns)	state:	Pembroke Dock(4), Devonport(9), Portsmouth(7), Woolwich(3), Deptford(1), Chatham(6), Kingston, CANADA(3)
2nd rate (90-98 guns)	state:	Devonport(9), Woolwich(5), Chatham(15), Pembroke Dock(9), Portsmouth(5), Deptford(3), Bombay, INDIA(4)
	private:	Milford Haven(1)
3rd rate (74-80 guns)	state:	Devonport(2), Pembroke Dock(6), Woolwich(11), Deptford(12), Portsmouth(5), Chatham(8), Bombay, INDIA(6)
	private:	Thames(43), Medway(7), Harwich(3), Hampshire(6) Devon(2), Milford Haven(1), Humber(1)
4th rate (50-60 guns)	state:	Sheerness(2), Pembroke Dock(4), Woolwich(5), Deptford(5), Portsmouth(3), Devonport(6), Chatham(2), Kingston, CANADA(2)
	private:	Thames(11), Medway(1)
5th rate (32-40 guns)	state:	Chatham(28), Deptford(19), Portsmouth(8), Woolwich(16) Sheerness(1), Pembroke Dock(17), Devonport(11) Bombay, INDIA(9), Kingston, CANADA(1)
	private:	Thames(33), Hampshire(26), Medway(13), Tyne(4), Humber(3), Harwich(2), Essex(1), Devon(1), Mersey(1), Milford Haven(1)
6th rate (20-28 guns)	state:	Pembroke Dock(11), Portsmouth(9), Chatham(5), Sheerness(4), Devonport(4), Woolwich(4), Cochin, INDIA(3)
	private:	Devon(13), Medway(6), Hampshire(6), Tyne(5), Chester(4), Ipswich(3), Dorset(2), Norfolk(3), Essex(2), Sussex(1)
sloop (10-18 guns)	state:	Portsmouth(32), Chatham(28), Pembroke Dock(21), Devonport(18), Woolwich(15), Deptford(9), Sheerness(2), Bombay, INDIA(19), Kingston, CANADA(2)
	private:	Thames(49), Devon(29), Kent(28), Hampshire(28), Medway(25), Ipswich(20), Norfolk(18), Essex(11), Tyne(6), Dorset(6), Humber(5), Cornwall(3), Harwich(2), Sussex(2), Bristol(1), Milford Haven(1), Berwick(1), BERMUDA(11), CANADA(3), Nova Scotia(2)

Note: [1]. excludes steam-powered vessels and sailing ships smaller than sloops.

in the sense that it represented the coalescence of new technologies in a variety of fields--ordnance, steam propulsion, metallurgy--which, when applied to that end, dramatically transformed weapons systems. Initially, the new military technology owed much to

product and process innovations in the civil field
(though not all, the development of machine tools in
the first place and the principle of interchangeable
parts in standard products in the second can be
ascribed to defence enterprises).[10] Latterly,
though, military and naval requirements grew to
become so specific that 'tools and equipment used to
make weapons gradually ceased to be suitable for
civil as well as military production'. The resultant special-purpose machinery increasingly took the
form of 'lumpy' capital plant which required continuous working--in the manner of follow-on defence
contracts--as the only feasible way of meeting amor-
tisation costs. Since the process technology was
specialised and not readily replicated in government
arsenals, it fell to an eclectic group of private
entrepreneurs to innovate the plant, oversee its
development and organise the means of production in
new MIEs. Usually, the MIEs began as branches of
existing civil enterprises in a cognate field of
heavy engineering: that way the civil and defence
businesses could complement each other in countering
alternating market downturns. This counter-cyclical
function is implicit in the entry of heavy engineers
into defence manufacture. All of the great MIE pioneer entrepreneurs--Armstrong, Vickers, Schneider
and Krupp--originally switched their foundries and
forges to arms manufacture when the first bloom in
railway expansion had dimmed. In due course, their
plant became vested with a defence speciality to
such an extent that they could not easily revert
much of their capacity to railway or structural
engineering work. By then, the firms were true MIEs
and, as such, seriously interested in influencing
the polity to ensure that defence contracts continued to be forthcoming. The early fusing of the
industrial and political interest into a prototype
military-industrial complex had its regional count-
erpart. Since the expensive new defence plant was
naturally set up alongside the mines and workshops
of the burgeoning industrial districts, lobbying in
favour of 'defence enclaves' in novel geographical
constituencies became a new force in the political
landscape. In the words of Pearton, the new MIE masters 'took care to interest politicians and officials active in the localities in which their factories were located, in the well-being of their
enterprises'.[11]

The promise of an emergent pork-barrel style of
politics was not the only facet of defence industry
inaugurated at this time. The necessity for keeping
operational the said MIEs with their new-found

expertise, meant that governments in the UK, France and Germany were amenable to pricing regimes that guaranteed profits to the entrepreneurs by way of domestic military contracts. In lieu of domestic orders and hidden subsidies, governments encouraged the firms to cultivate export markets. Indeed, for much of the time the MIEs depended for their very survival on export orders: until the later decades of the century Krupp was a larger gun supplier to the Tsar than he was to the Prussian ordnance department whereas Armstrong innovated a new class of warship, the protected cruiser, on the strength of a string of orders from overseas customers ranging from Chile to China. A primitive kind of offset policy was also much in evidence. For example, the UK Admiralty refused to purchase torpedoes from the innovator Whitehead unless he arranged a British production facility in addition to his main plant at Fiume in Austria-Hungary (now Rijeka, Yugoslavia). The offset for production contracts with a foreign MIE materialised in 1891 in the guise of the Weymouth torpedo factory (later to be absorbed--along with the parent organisation--by Vickers and Armstrong at Foreign Office instigation). The bail-out expedient can also be traced to this formative era. No less a personage than Krupp was saved from bankruptcy after over-extending his operations in 1874 by a timely infusion of capital orchestrated by the Prussian State Bank. Finally, collaboration between MIEs with the object of either sharing technical capabilities or maintaining adequate prices was a feature of the times and, as such, a precursor of the international ventures and the prime/subcontractor relations which so typify the current defence-industrial scene. The most famous of these so-called coalitions was the Steel Manufacturers' Nickel Syndicate drawing together the armour producers of the USA, UK, Germany and France. Many of these groupings represented multinational initiatives on the part of established MIEs and the beginnings of import-substitution-industrialisation elsewhere.

Endemic to the early phase of MIE formation was the retreat of government from much of the innovative activity in armaments manufacture. In part this was a reaction to the abysmal performance of state undertakings during the Crimean War. The giants of 19th century gun design, William Armstrong and Joseph Whitworth, first applied scientific principles at this juncture, stepping into the breach left by the inadequate arsenal system. Concurrently, John Brown and Charles Palmer, amongst others, turned their formidable metallurgical skills to the

new naval contingency of armoured defences. For a
time, the Royal Arsenal at Woolwich was confronted
with competition from private suppliers of guns of
undoubted superior quality, although in a postwar
fit of protectionism it gained some monopolistic
respite until its lackadaisical design efforts
forced a reluctant government to reintroduce gun
orders to private firms after 1880. The Royal Small
Arms Factory at Enfield Lock, which had replaced the
Tower of London in 1811 as the chief supplier of
British small arms, escaped such ignominy because it
was revamped in the 1850s through the drastic measure of importing US personnel, management and production methods. This, an early example of technology transfer from one country to another,
safeguarded the state arsenal system, but private
suppliers of rifles and revolvers acted as a supplement to government production from thenceforth.
Interestingly, UK private manufacturers of small
arms were allegedly disadvantaged in terms of process technology relative to their highly innovative
US counterparts--early advocates of mass-produced
weapons with interchangeable parts--because of the
British contracting system. The UK predilection for
awarding the lion's share of military contracts to
the state system, with the corollary of penny-packet
orders at infrequent times to the contractors, had
only one possible outcome: the enforced devotion of
the said contractors to labour-intensive batch production largely for civil outlets. A combination of
bought-in US technology, licence-production of emi-
nent foreign designs, and some incremental innovation, enabled the Enfield arsenal to stay abreast of
the private firms.[12] The same could not be said of
the state system in the heavy engineering aspects of
defence production. For one thing, European innovations in the fields of ballistics, metallurgical
product and process technology, and marine engineering and ship construction were unsurpassed precisely
because of their transferability in the early stages
from civil to defence industry by entrepreneur-inno-
vators with the foresight to recognise the potential
of manufacturing for defence markets. The short
order wherein firms shifted from the conception
stage through to fully-fledged mature MIEs occupies
much of the remainder of the chapter. However, a
brief overview of what was entailed in the technical
and organisational steps undergone by each of the
main defence industry constituents--ordnance, armour
and shipbuilding--is a necessary preliminary.

Ordnance

The conception stage for ordnance revolved round the rifled artillery piece which subsequently stabilised on breech-loading mechanisms. Innovators in this area were Armstrong and Whitworth, the former establishing a tentative ordnance R & D centre at Elswick (Newcastle) in 1847 while the latter began to lay down gun-making plant at his Openshaw (Manchester) engineering works after 1854. The Crimean War and the subsequent rivalry of the 'ironclad' naval race of the early-1860s conspired to foster official interest in the innovations, ensuring that the 'birth' stage of experimentation and the 'childhood' stage of MIE formation were consolidated. By the 1870s metallurgical improvements had enabled the gun-makers to produce a piece sufficiently strong to overcome the inherent weaknesses of a breech opening, and henceforward breech-loading cannon were demonstrably superior to muzzle loaders. The adoption by the Royal Navy of the breech loader and, in particular, Armstrong's quick-firing 4.7-inch gun after 1886 and 6-inch gun after 1890, set the seal of approval on private manufacture of artillery.[13] The Royal Arsenal was steadily relegated to secondary standing as the state formally spurred the entry of private firms into gun-making; most notably, Vickers in 1888. In fact, the decision taken by the government in 1884 to rely on private industry for heavy gun forgings capacity was of considerable import; for, not only did it lead to the entry of Vickers, Cammells, Beardmores and other steel-makers into the artillery business, but it enticed the same heavy engineering firms into becoming general defence contractors in the decades following the 1889 Naval Defence Act and the huge rearmament programme that the legislation engendered. The climax of artillery production in peacetime occurred in the first decades of this century when guns and gun mountings were integral--and vital--components of the battleship weapon system. At that juncture, Vickers and Armstrong Whitworth had been joined by the Coventry Ordnance Works (a MIE formed in 1907 and jointly owned by John Brown, Fairfield and Cammell Laird) and Beardmore in the UK, the German scene was dominated by Krupp while that of France was commanded by Schneider and the USA by Bethlehem.
Innovation in light ordnance centred on the machine gun. Monopolised by US innovators, the machine gun first received practical expression in Europe with the realisation of 'childhood'-stage enterprises, transhipped to the Old World in hopes

of fulfilling greater market demand. Benjamin
Berkeley Hotchkiss, for one, emigrated in 1867 to
France where he had his machine gun adopted by the
army and his company was commissioned to design
subsequent guns (incidentally, he also succeeded in
selling licence rights to Armstrong in 1884 so that
the gun could be made at Elswick and in 1887 he was
induced to start a Connecticut factory on the
strength of an order for ninety guns placed by the
US Navy). For another, Hiram Maxim quit his native
Maine and opted for British nationality, receiving a
knighthood for his efforts in perfecting the machine
gun. His gun was adopted by the British Army in 1891
and his factory for making them (a co-operative venture with the Swede, Nordenfelt) was absorbed by
Vickers in 1897. Maxim's gun designs formed the
inventory of all the world's major armies by 1914
with the exception of the French and Japanese who
had put their faith in Hotchkiss.[14] By that time,
yet another US design was taken up by Europeans for
domestic manufacture; in this case the BSA firm
turned its Birmingham factory over to the manufacture of a machine gun formulated by Isaac Lewis of
the Automatic Arms Company of Buffalo, New York.
Lewis was also instrumental in establishing the
Armes Automatique Lewis enterprise in Belgium in
1913 for the purpose of making the gun for the Belgian army. It is a matter of fact that US innovators
inspired the arms industry throughout Europe. The
automatic firing principle when applied to small
arms—the pistol and then the rifle—owed much to
the pioneering efforts of John Browning. His early
work in the machine-gun field resulted in a gas-op-
erated mechanism that was incorporated into guns
built by the famous Colt firm in the USA. Subsequent
work in small arms led to the licensing of automatic
pistols to the FN enterprise in Belgium. This latter
firm had been established in 1889 to make German
Mauser rifles under licence: its adoption of Browning patents in 1896 gave it a commanding position in
European small arms manufacture.

Machine guns, like heavy artillery but unlike
small arms, were essentially a preoccupation of private MIEs—and for the same reason: design resided
in the innovative endeavours of private entrepreneurs rather than in the official staffs of government arsenals. Thus, maturity for the machine-gun
manufacturers derived from technology transfer, in
the main from the USA, combined with a large and
expanding market in Europe where rearmamental instability was the rage. The production of machine guns
was undertaken by specialist gun firms on the one

side (small arms manufacturers effecting related diversification into machine gun technology) and heavy engineering enterprises on the other (acquiring a wide-range of armaments capabilities including both heavy and light ordnance manufacture). In conjunction, maturity of ordnance production executed by the two types of suppliers had important repercussions for ammunition supply. All the manufacturers of guns were equally conversant with ammunition production, but the increasing complexity of the chemistry intrinsic to shell filling explosives led to the entry of specialist undertakings. In the UK, for instance, armour piercing shot became the preserve of special-steels enterprises Thomas Firth and Hadfields of Sheffield. Shell-making, meanwhile, was pursued by those two firms along with the Royal Arsenal, the clutch of general defence contractors—Armstrong Whitworth, Vickers, Beardmore and Cammell Laird—as well as a specialist shell-maker, the Projectile Engineering Company, founded in 1902. Small arms manufacturing was carried out by the Enfield arsenal and a number of non-ferrous metal manufacturers located around Birmingham.[15] Somewhat perversely, the BSA firm had divested itself of ammunition production in 1897 with the sale of its plant to the Nobel explosives organisation (whereupon it was renamed the Birmingham Metals & Munitions Company). In a similar fashion, an explosives company in America, Du Pont, was to enter the military ammunition market in a big way (in 1935) through the purchase of a small arms interest; in this case the entire assets of Remington, a company founded at Ilion, New York, in 1816.

Armour

Creativity in armour technology derived from both sides of the Atlantic. The original impetus for it had arisen in the Crimean War when armoured floating gun batteries were put into use. Subsequently adopted for warships, the 'ironclad' era was inaugurated by the French wooden-hulled ironclad 'Gloire' in 1859 and the British all-iron broadside ships 'Warrior' and 'Black Prince' of 1860-1. Significantly, the French ship was a product of a state arsenal (Toulon) whereas the British ships emanated from private shipyards (Ditchburn & Mare on the Thames and Napier's of Glasgow). The latter circumstance was a consequence of the mutually beneficial co-operation between the Admiralty's chief constructor, Isaac Watts, and the private engineering

143

genius, John Scott Russell. The operations of iron-clad naval monitors in the American Civil War soon endorsed the urgency of acquisition of armoured ships by the onlooking naval powers. Two of the best-equipped private firms able to fulfil that demand were shipbuilders which had invested in iron armour-plate facilities. The Thames Iron Works of Blackwall had, by 1863, achieved an annual capacity of 25,000 tons of ironclad ships, and its ancillary equipment included three rolling mills for manufac-turing angle irons and wrought iron plates from scrap as well as seven large steam hammers for shap-ing the plates and forgings. The neighbouring Mill-wall Ironworks boasted an armour-plate mill worth in the region of £100,000 and capable of turning out 15,000 tons of armour plate per year.[16] However, the writing was on the wall for the Thames contractors even at this relatively early date. Already the North of England was beginning to rival London in forgings and plate manufacture: the Mersey Steel & Iron Works of Liverpool, John Brown's Atlas Works at Sheffield and Samuel Beale's Parkgate Ironworks near Rotherham were all auguries of the new wave of steel specialists which would come to dominate armour mak-ing. Inability to advance the state-of-the-art doomed the Thames contractors to eventual extinc-tion.

After a busy decade or more of building iron-armoured ships, innovations in steel technology allowed for the introduction of compound armour plate after 1877; some of the best of which was pro-duced by Cammells and Browns in the UK. Compound armour, as its name implies, was a combination of wrought iron and steel, and was instantly adopted by the Royal Navy. It, rather than pure steel plate, was adopted because steel suffered from the flaw of cracking. Nevertheless, steel was lighter and tougher than iron and, after several years of exper-imentation, Schneider's all-steel plate was taken up by the French and the budding naval power of Italy (which immediately set about erecting a plant spe-cifically to produce it). In the final decade of the century, technological advances were such as to allow for the full adoption of steel plate. The ear-lier problem of a tendency to cracking was progres-sively eliminated. Moreover, the innate resistance of steel was augmented through a process of armour hardening effected by means of chilling the heated surface of the steel by high-pressure water jets. This case hardening was greatly improved by the US innovator, H. A. Harvey, who introduced a procedure in 1891 whereby plates could be case hardened

without the need for a wrought iron backing. At one
fell swoop, steel was rid of its previous structural
drawback while compound armour was rendered obsolete. To boost its resistance properties even further, steel armour was converted into an alloy in
1889 when Schneider added nickel, and in 1893-4
Krupp capped it all by innovating a special heat
treatment process for making nickel-chrome steel
even harder than that subject to case hardening
(i.e. Krupp's decremental hardening method). The
world's armour producers responded by taking out
licences from both Harvey and Krupp in order to
remain abreast of current technology. By 1914
roughly comparable armour was being produced in Germany by Krupp and Dillengen; in the UK by Vickers,
John Brown, Hadfield, Cammell Laird, Beardmore and
Armstrong Whitworth;in France by a bevy of firms but
especially Schneider and the state-owned Forges
Nationales de las Chaussade; in Italy by Terni-Vick-
ers and Ansaldo-Armstrong; in Russia by Nicopol-Mar-
ionpol; and in Japan by the state's Kure dockyard
and Wakamatsu Iron Works.[17] For its part, the USA
supported three private armour-making enterprises:
the Bethlehem, Midvale and Carnegie steel companies
of South Bethlehem, Philadelphia and Munhall, Pennsylvania, respectively. The first had been induced
to enter armour-plate production in 1887 on receipt
of an order for 14,000 tons from the US Navy. It
responded by purchasing, lock, stock and barrel, an
armour forging plant—at government expense ($4.5
million)—from Whitworth in England. Wishing to
establish a second-source, the US Government gave
Carnegie (US Steel after 1901) a contract for 6,000
tons of armour in 1892 on condition that a forgings
plant was erected at the Homestead Works. Midvale
had concentrated on gun forgings for the US Navy,
but the Philadelphia company entered the armour
business in 1903 to counter the inroads made by
Bethlehem in the gun forgings area.

One firm to prosper amply from the armour business was Cammells. The pioneer entrepreneur Charles
Cammell had started a file manufacturing operation
in Sheffield in 1837, expanded into the forgings
trade in 1845 with the opening of the Cyclops Works,
and confirmed the 'childhood' stage of corporate
development with the formal creation of Charles Cammell & Company in 1854. The firm's principal product
in its first flowering was wrought iron plate made
from cold blast pig iron, but after 1861 attention
shifted to the Bessemer steel-making process. Cammell was an early advocate of compound armour, rolling the plate at the Cyclops Works. Acknowledging

145

the need for a counter-cyclical capability, the firm
built a new steelworks at Grimethorpe to make use of
Siemens open-hearth furnaces for the production of
railway tyres and axles, acquired the Penistone
Works to make steel rails using Bessemer plant and,
in 1870, set up a steel wheel-making plant at Dron-
field. Unacceptably high transport costs incurred in
working this last facility led to its complete closure and dismantling in 1883 for transfer to Workington, Cumbria. Not only was the new location on
the coast and therefore easily accessible to marine
transportation, but it was situated close to the
company's newly-acquired Eskett Park, Mowbray, Park-
house and Frizington Parks hematite iron ore
mines.[18] Workington became the centre of the company's civil work (complemented in 1895 by the purchase of the adjacent Solway Works in Maryport for
merchant iron materials), leaving the Sheffield
operations to focus on armour plate and gun forgings. Profitable defence orders at the turn of the
century persuaded the firm to extend its defence
business and in 1903 Cammells acquired the Merseyside (Birkenhead) shipbuilder of Laird Brothers. The
new Cammell Laird & Company had its capital structure boosted from £1.555 million to £2.5 million and
was poised to enjoy the benefits flowing from the
vigorous battleship building programme then under
way.

Also hailing from Sheffield was the firm of
John Brown's and this enterprise trod a similar path
to Cammells. Brown had erected a rolling mill in
1859 able to manufacture five-ton armour plate and,
by 1867, fully three-quarters of the ironclads in
the Royal Navy used plate deriving from his Atlas
Works. As a result, Browns had become a public company in 1864 (capitalised for £1 million) and had
witnessed a meteoric rise in employment, from 200 in
1857 to 4,000 ten years later. Diminishing Admiralty
demand for iron armour in the late-1870s stimulated
the firm to experiment with compound armour and,
along with Cammells, it came to dominate this product line in the 1880s. The firm's defence-dependence
was furthered after 1886 with the erection of a
large forging press specially designed for the manufacture of heavy forgings for guns and marine shafting for warships. Browns' own R & D efforts pre-
empted Harvey's process of armour hardening and the
firm thereby managed to escape the payment of royal-
ties to the American innovator. On the installation
of a 10,000-ton armour-plate press made by Whitworth
and the obtaining of Krupp licences, the firm was
well placed to compete for armour orders during the

heightened naval expansion of the 1890s and after. It did not rest on its laurels, however. Like Cammells, it decided to purchase a shipbuilder well equipped to participate in warship construction. Its choice was the Clydebank Engineering & Shipbuilding Company (J & G Thomson) which, on takeover in 1899, raised Browns capital from £1.25 million to £2.5 million. Three years later, John Brown & Company greatly strengthened its ordnance interests by buying into Thomas Firth & Sons. Firths had began as makers of files, saws and edge tools in Sheffield, but in 1852 the company forged its first steel gun at the Claywheel Forge. By 1870 its revamped Norfolk Works made it 'probably the best gun steel and forging firm in the world'.[19] It was also a leading manufacturer of shell projectiles. The half share of Browns in the Coventry Ordnance Works merely endorsed its role as a leading ordnance contractor. Thus, by the eve of World War 1, the firm was producing armour plate at the Atlas Works for installation on warships built at Clydebank and for use on the heavy gun mountings made by the Coventry Ordnance Works and Firths. The company's interests (along with Vickers) in the Spanish dockyards at Ferrol and Cartagena provided it with a further outlet for its armour and, just to ensure a civil market alternative, its controlling interest from 1906 in Belfast shipbuilder Harland & Wolff guaranteed Browns' forgings and castings access to the liner trades and other merchant shipbuilding.

Warships

Rather surprisingly in view of later events, the first private MIEs extensively engaged in warship construction were the Thames firms of Penn and Maudslay. Building on their pioneering skills in steam propulsion, these marine engineers supplied the engines for a vast fleet of 150 gunboats ordered in the Crimean War and they gave a glimmering of early mass-production techniques.[20] Their example was quickly emulated by other marine engineers able and willing to fill a void in Royal Dockyard capabilities. John Elder, for example, who built his first marine engine in 1853, decided in 1860 to enter shipbuilding and four years later laid out a shipyard and engine works at Fairfield (Govan) on the Clyde. Reconstructed as Fairfield Shipbuilding & Engineering Company in 1886, the Glasgow firm was to become a major warship supplier.[21] Yet, despite deficiencies in propulsion machinery, armour and gun capabilities--all of which had to be bought-in from outside, mostly private, sources--the Royal Dockyards maintained a healthy competitive edge over private shipbuilders for many years. Indeed, as late as 1905-6, that augury of modern capital ships and the vessel after which a whole fleet of ships was named, the 'Dreadnought', was built at Portsmouth Dockyard as this yard was reckoned to be the fastest builder in the country.[22] On the whole, the Admiralty preferred to use its own facilities for hull construction and, as in the Napoleonic era, imposed a rough division of labour in ship procurement.

As can be elicited from Table 4.2, the majority of the battleship classes built in the ironclad and steel armour period were obtained from the Royal Dockyards and, in particular, Portsmouth, Chatham, Pembroke Dock and Devonport. The residual 42 per cent of such ships derived from private yards, but in contradistinction to the pre-steam navy, came overwhelmingly from the North of England and the Clyde. The intermediate class of cruiser-type vessels was more evenly shared between government and private producers: indeed, the latter enjoyed a slight advantage in that commercial builders (predominantly on the Clyde) supplied 52 per cent of them. Furthermore, the Admiralty brought in additional producers, using yards on the Humber (Earle's at Hull), the Wear and the Tees. Moreover, it pressed into service some of its smaller dockyards, most notably Sheerness and, prior to their closure in 1869, Woolwich and Deptford. The epitome of imperial policing, the sloop-cum-gunboat, enjoyed widespread distribution of supply; although the Royal Dockyards built 51 per cent of the 326 ships in question (including a solitary vessel produced by Malta Dockyard, the only overseas warship source for the Royal Navy during this period). Again, the Admiralty tried an expanded list of commercial shipbuilders, encompassing yards in Hampshire and Belfast as well as the aforementioned districts. In marked contrast, no destroyers were built in state yards: a reflection of the expertise vested in their private innovators, Yarrow and Thornycroft, who initially built on the Thames before transferring to the Clyde and Hampshire (Southampton) respectively. Other private companies availed themselves of the opportunity to enter the innovative destroyer field---Lairds on Merseyside, Palmers on the Tyne and Whites on the Isle of Wight were not remiss in this respect. Following destroyers at the turn of the century was another warship innovation, the submarine, and here Vickers--partly through its obtaining

Table 4.2 : British Warship Contracts, 1860-1914

Ship type	Distribution (number): state	private
Battleships, battlecruisers, ironclads, ships of the line, turret ships	Portsmouth (25) Chatham (23), Woolwich (3) Pembroke Dock (17) Devonport (16)	Clyde (18) Tyne (16) Thames (12) Mersey (8) Barrow (7)
Cruisers, frigates, corvettes, monitors, rams	Chatham (36), Sheerness (16) Pembroke Dock (33) Devonport (25) Portsmouth (22) Woolwich (6), Deptford (6)	Clyde (72) Tyne (30), Wear (1), Tees (1) Barrow (22) Thames (16) Mersey (8) Humber (7)
Sloops, gunboats, large torpedo boats	Pembroke Dock (40) Devonport (38) Sheerness (27), Chatham (23) Portsmouth (19) Deptford (17), Woolwich (1) MALTA (1)	Thames (35) Tyne (34), Wear (4), Tees (3) Clyde (27), Forth (1) Hants (19), Kent (1) Mersey (18) Belfast (4), Humber (3), Milford (1) Barrow (10)
Torpedo-boat destroyers		Tyne (81), Clyde (71), Hants (39) Mersey (36), Thames (33), Barrow (9) Wear (6), Humber (4)
Submarines	Chatham (15)	Barrow (77) Tyne (1), Clyde (1)

of US Electric Boat licences--became a monopoly supplier. Consequently, Barrow produced all the British
boats until the Admiralty brought in, first, Chatham
Dockyard in 1907 to allow it to keep up with submarine design and, secondly, private producers on the
Tyne (Armstrong) and Clyde (Scotts) in order to
diversify the sources of supply. In general, then,
the Admiralty attempted to retain in its own hands
production capacity for the larger warships and was
prepared to accept private production of smaller
ships, especially those like destroyers and submarines which were the offspring of private innovation.

The 1889 Naval Defence Act had put paid to the
simplicity of this division, however. As a result of
war alarms, the two-power standard was adopted
whereby the Royal Navy was to be of a size equal to
the combined strengths of the next two naval powers.
To that end, 70 warships worth £21.5 million were
ordered, and this was followed in 1893 by the
Spencer programme which sanctioned a further spate
of orders, including 82 destroyers. Consequently,
the expenditure on naval shipbuilding (excluding
ships' armament) leapt up from £2.6 million in
1888-9, to £3.6 million in the following year, £6.8
million in 1890-1, £6.3 million in 1891-2 and £4.7
million in 1892-3.[23] The consistently higher level
of warship demand effected after the 1889 legislation required shipbuilding capacity far in excess of
that available in Royal Dockyards and could only be
met by co-opting the commercial shipbuilding industry into the scheme. Some private firms with previous warship experience immediately took advantage of
the chance to profit from government orders and
engaged in 'buy-ins', that is, they offered tenders
lower than cost in order to gain contracts. Palmers
of Jarrow, for one, lost heavily on two battleships
which the firm had taken on, at low bids, simply to
counter the effects of recession in merchant shipbuilding. The firm admitted to a record loss of
£11,000 in 1891, opted to recommence merchant shipbuilding at Howden rather than concentrate on naval
orders, and sold its interest in the Bilbao naval
yard to the Spanish Government. Yet it was still
left with a new ordnance plant empty of work and
showed an even larger loss of £22,100 in 1892. The
conversion of the ordnance plant to bicycle manufacture and a more realistic approach to defence tendering enabled the company to endure: by the end of
the 1890s it was in a profitable situation again,
making, for example, £107,074 profit in 1901 on a
capitalisation of £883,145.[24] Less fortunate was the

firm of Samuda Brothers of Cubitt Town, London. Its
contract for two cruisers under the 1889 programme
only served to overstretch resources and bring on
the firm's failure.
    Casualties notwithstanding, the Naval Defence
Act transformed commercial shipbuilding and confirmed the rise of private MIEs. Some, such as
marine engineers Hawthorn Leslie of Newcastle, had
been unsuccessful merchant shipbuilders that were
electrified by naval possibilities. Specialising in
the new destroyer type, Hawthorns began to make satisfactory profits after 1898: the outcome of combined sales of hulls and main engines to the Admiralty.[25] Indeed, small specialist shipbuilders of
the likes of Yarrow, Thornycroft and J. S. White
owed their entire position as successful firms to
Admiralty appreciation of their skills as innovators
of small warships, engines and boilers. Most of the
firms enjoined to enter naval construction, however,
focused on the advantages to be gained from building
battleships and cruisers. In so far as battleships
are concerned, only seven had been built in private
yards between 1869 and 1890 in comparison with the
23 produced by the Royal Dockyards. For the period
spanning the years 1893 to 1913, the situation moved
towards balance: private contractors delivered 36
ships as against the 40 emanating from the dockyards.[26] Indeed, in the decade prior to World War 1
which covers the inception of the 'dreadnought' all-
big gun battleship, the situation was reversed, with
the private MIEs exceeding the battleship production
of the state facilities.
    The transition from dockyard dominance to pri-
vate MIE pre-eminence is traceable through Figures
4.1 and 4.2. The first evinces the contrast between
the era immediately prior to the Naval Defence Act
and that which succeeded it. In the 1880s, the dockyards built 79 per cent of capital ship tonnage,
leaving the rest to be provided by the Thames Iron
Works and Armstrong. The dockyards still managed to
build the bulk of a much-expanded tonnage in the
post-1889 decade—72 per cent—but now the private
sector had witnessed several new entries; namely,
Laird, Thomson, Palmers and Vickers (Armstrong was
diverted to building warship tonnage for export).
Figure 4.2 indicates that though the 'dreadnought'
innovation was led by two dockyards, Portsmouth and
Devonport, the weight of tonnage—62 per cent of the
876,800 tons delivered between 1905 and
1914—derived in the main from private yards. Of the
'dreadnought' builders, only Palmers, Scotts and the
Thames Iron Works were arguably mainly occupied as

Figure 4.1 : Capital Ship Production, 1880-99

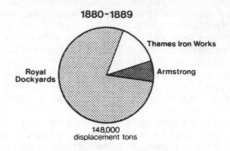

1880-1889

Thames Iron Works

Royal
Dockyards

Armstrong

148,000
displacement tons

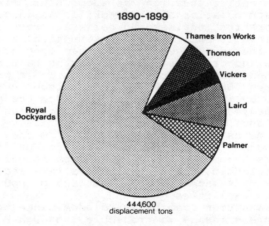

1890-1899

Thames Iron Works

Thomson

Vickers

Royal
Dockyards

Laird

Palmer

444,600
displacement tons

shipbuilders (and the Thames firm had expired by
1912); the others--Armstrong Whitworth, Vickers,
Browns, Fairfield, Beardmore and Cammell Laird--were
more diversified defence contractors. In part this
was occasioned by the fact that diversification was
both advantageous and profitable. Clearly, it
offered security of earnings through access to multiple markets. Also, it gave a MIE control of many
of the constituent parts of a battleship weapon system. For example, the gun mountings of a 'dreadnought' comprised 21 per cent of ship costs; a far
higher figure than the nine per cent of costs that
they amounted to in a battleship of the Spencer programme.[27] Obviously, there was sufficient incentive
here to induce ordnance firms to integrate into
shipbuilding. Similarly, as the new types of capital
ships were larger than their predecessors, the

absolute value of armour built into them mounted
correspondingly, whereupon armour producers felt
encouraged to acquire shipbuilding outlets for their
product. By all events, weapons technology was drawing together the manufacturers of guns and platforms, a phenomenon to which we now turn.

The Drive to Integrated Operations

The impetus for the broad-based MIE with capabilities in guns, armour and warships arose initially in
the USA. It was, in fact, the result of a deliberate
act of government policy. Intent on creating a
respectable 'blue-water' fleet in lieu of a mere
coast defence force, the US Navy initiated its first
steel warship programme, the so-called ABCD scheme,
in 1883. Immediately, structural defects in US
defence industry were exposed, not least of them
being the total inability of state dockyards and
arsenals to meet the challenge. Having little
recourse but to use private firms, the government
virtually founded an arms complex from scratch,
although not without considerable difficulty. For
example, the first naval programme exacted a heavy
cost, including the bankruptcy of shipbuilder Roach
who had all the hull contracts. Moreover, Midvale
Steel was only capable of forging guns up to 6-inch
calibre and the forgings for larger weapons had to
be imported from Cammells and Whitworth.[28] In order
to put the situation to rights, the US Navy systematically encouraged new plant for warships, guns and
armour. As for the first, the Newport News Shipbuilding & Drydock Company stands as the supreme
example. Founded by Collis Huntington, the owner of
the Chesapeake & Ohio Railroad, the company opened
its first drydock at the mouth of the James River in
1889. Within three years the firm had obtained its
first large navy contract, and by 1907, when President Roosevelt despatched his 'Great White Fleet' on
a world tour, no fewer than seven of the 16 assembled battleships had emanated from Newport News.[29]
In the 1890s, then, the firm was well-placed to
undertake the construction of US warships and was
joined in this endeavour by Cramp in Philadelphia,
the Union Iron Works in San Francisco, the Bath Iron
Works of Bath, Maine, and the Moran Brothers with
their shipyard in Seattle, as well as by some of the
belatedly refurbished government-owned Navy Yards.
Both state and private facilities were pressed into
gun supply too. Gun factories were set up at the

Figure 4.2 : 'Dreadnought' Tonnage by MIE, 1905-14

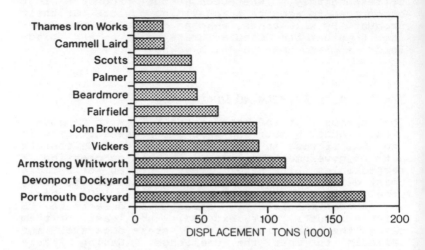

Army's Watervliet Arsenal in Troy, New York, and the Navy's Washington Yard; whereas private interests were represented by the American Ordnance Company (of Washington) and the Bethlehem Steel Company. By the early-1900s Midvale Steel had also become a fully-fledged ordnance producer. Midvale and Bethlehem were also prominent, of course, in armour-plate manufacture. Actively encouraged by the US Navy, both firms aimed at becoming integrated arms producers. Midvale was to go on to acquire the Remington small arms manufacturer while Bethlehem entered shipbuilding in a serious way. At first rebuffed with its 1902-3 involvement in the US Shipbuilding Company, Bethlehem had by 1914 control of the Union Iron Works in San Francisco and the Fore River shipyard at Quincy, Massachusetts. Between them, these yards built two 'dreadnoughts', five destroyers and 22 submarines from 1906 to 1914.

Attempts at fomenting integrated operations in the UK derived from two stimuli: first, a continuation of the market-induced process of vertical integration and, secondly--and as remarked previously--in response to the post-1889 era of naval expansionism. Palmers serves as a classic instance of the former case. Its fame followed from the innovation of iron-built steam colliers in 1852 and the firm's expertise in iron shipbuilding was unassailable. It erected four blast furnaces at Jarrow to serve its shipyard there, and bought coke ovens

across the Tyne at Wallsend. In 1864 it progressed
to rolling mills and opened a supplementary shipyard
at Howden. It was during this period that Palmers
began to enter defence markets by applying armour
plate technology learned from the Parkgate Ironworks
at Rotherham. Expansion of ironmaking induced it to
open its own iron ore mines at Port Mulgrave, ten
miles north of Whitby. Palmer's Shipbuilding & Iron
Company was, by 1883, turning out an average of
57,000 tons of merchant shipping each year, had a
capacity for 50,000 tons of finished iron, and was
employing 7,500 workers. As we have seen, naval
rearmament resulted in mixed fortunes for the firm;
but by 1911 it had reserved the main Jarrow establishment for warship work while its merchant shipbuilding was concentrated on the recently-acquired
Hebburn yard of Robert Stephenson & Company.[30] Thus,
Palmers was essentially a shipbuilder which had
integrated backwards through blast furnaces to iron
ore and coal mining, and forwards into battleship
construction along with its armour plate. The second
rationale for integration, meanwhile, was far more
pervasive, climaxing in the decade after 1897. The
chief events incident to its creation are summarised
in Figure 4.3. Triggered by the merger of Armstrong
and Whitworth, together with the purchase of the
Barrow shipyard and ordnance works by Vickers--both
in 1897--the next major move was the purchase of the
Clydebank shipyard by Browns in 1899; followed in
short order by the effective merger of Beardmore
into Vickers (1902), the merger of Cammells with the
Birkenhead shipyard of Laird (1903) and culminating
in the foundation in 1907 of the Coventry Ordnance
Works by Fairfield, Cammell Laird and Browns. To put
the rise of the general defence contractor into perspective requires a brief elucidation of the principal players.

Armstrong Whitworth.
W. G. Armstrong bought 5.5 acres of land at Elswick
in 1847 to set up a factory for making hydraulic
cranes. Motivated by the exigencies of the Crimean
War, he invented breech-loading cannon with rifled
barrels and patented the invention in 1859. In the
same year, he established the Elswick Ordnance Company alongside the crane works. Revisionist thinking
at the War Office in 1863 led to the military
reverting to muzzle-loading cannon and Armstrong was
obliged to cultivate export orders for his guns
(until 1877 when the British Government gave him
contracts again) and diversify into blast furnace

Figure 4.3 : Creation of Integrated MIEs

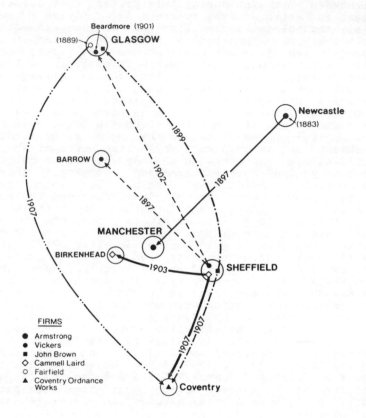

activities (until 1900). In 1882 Armstrong amalga-
mated with Mitchell & Swan, a Tyneside merchant
shipbuilder, in order to enter shipbuilding. In the
following year, Armstrong-Mitchell opened a new warship-building yard at Elswick (congruent with the
firm's innovation of the protected cruiser), reserv-
ing the Low Walker yard (Mitchell's)   for merchant
shipbuilding. By 1895 the Elswick complex alone
embraced nine acres of works producing hydraulic
machinery (employing 1,500), upwards of 40 acres for
the gun factory (employing 3,500), and 16 acres for
the warship yard (employing about 6,000). Two years
later, the firm absorbed Sir Joseph Whitworth & Company and gained access to the sprawling Openshaw

Works. Restyled Sir W. G. Armstrong, Whitworth &
Company, the merged firm employed in the region of
25,000 men and was self-sufficient in armour making.
From this year up until World War 1, around 90 per
cent of Armstrong Whitworth's resources were devoted
to defence production.[31] The firm entered that war
resplendent with steelworks at Openshaw; gun factories at Openshaw, Elswick and Pozzuoli, Italy;
explosives and armaments factories on the Thames and
at Alexandria, Dumbartonshire, and was about to
undertake military aircraft production as well as
opening a 'greenfield' Walker shipyard to supersede
the Elswick naval yard. For good measure, it also
ran a shipyard in Italy from 1904, in conjunction
with Ansaldo. A minimal counter-cyclical capability
was retained through interests in foundry steel (in
Canada), motor car manufacture and merchant shipbuilding.

Vickers.

Later to become synonymous with arms production,
Vickers had turned its Sheffield workshops to the
manufacture of armour and heavy guns as recounted
above. Ambitious to fill the gaps in its defence
portfolio, it simultaneously acquired the Naval Construction & Armaments Company of Barrow and Maxim
Nordenfelts of Crayford in 1897, whereupon it gained
a warship construction and arming complex and a
machine-gun factory. Five years later, the company
purchased a 50 per cent stake in William Beardmore &
Company. In so doing, the 'merger ensured for both
firms the light armour plates made by Vickers and
the heavy armour in which Beardmore specialised, as
well as allowing them to specialise in their three
shipyards at Barrow, Dalmuir and Govan'.[32] Less significant activities at this time included the establishment of the Wolseley Tool & Motor Car Company
(1901) with 'half an eye on military possibilities',
not to speak of a string of overseas defence interests (Figure 4.4). Chronologically, the first was a
joint venture of 1897 with the Spanish Government
(Placentia de las Armas) for the formation of an
ordnance works: eleven years later Vickers was to
join with Armstrong Whitworth and John Brown in a
24.5 per cent involvement in La Sociedad Espanola de
Construccion Naval (SECN). After obtaining submarine
licences from the US Electric Boat Company in 1900,
Vickers proceeded to purchase a substantial minority
holding in the innovating submarine builder in 1902.
Its Italian ventures focused on a part share of the
Vickers-Terni gun works at La Spezia (from 1905)

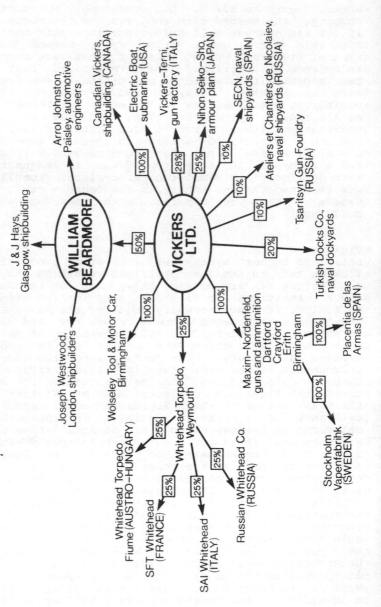

Figure 4.4 : The Vickers-Beardmore MIE Combine, 1914

WILLIAM BEARDMORE

- J & J Hays, Glasgow, shipbuilding
- Arrol Johnston, Paisley, automotive engineers
- Joseph Westwood, London, shipbuilders

VICKERS LTD.

- 50% → WILLIAM BEARDMORE
- 100% → Canadian Vickers, shipbuilding (CANADA)
- 100% → Electric Boat, submarine (USA)
- 28% → Vickers–Terni, gun factory (ITALY)
- 25% → Nihon Seiko-Sho, armour plant (JAPAN)
- 10% → SECN, naval shipyards (SPAIN)
- 10% → Ateliers et Chantiers de Nicolaiev, naval shipyards (RUSSIA)
- 10% → Tsaritsyn Gun Foundry (RUSSIA)
- 20% → Turkish Docks Co., naval dockyards
- 100% → Wolseley Tool & Motor Car, Birmingham
- 25% → Whitehead Torpedo, Weymouth
- 100% → Maxim–Nordenfeld, guns and ammunition Dartford Crayford Erith Birmingham
- 100% → Placentia de las Armas (SPAIN)
- 100% → Stockholm Vapenfabrihk (SWEDEN)

Whitehead Torpedo, Weymouth

- 25% → Whitehead Torpedo Fiume (AUSTRO–HUNGARY)
- 25% → SFT Whitehead (FRANCE)
- 25% → SAI Whitehead (ITALY)
- 25% → Russian Whitehead Co. (RUSSIA)

while its purchase, along with Armstrong, of the Whitehead Torpedo Works in 1907, gave it a stake in factories at Fiume (Austria-Hungary) and Weymouth (England). These two firms also co-operated with Mitsui in the formation of a Japanese armour-making plant in 1907 and were poised to revamp Turkish state arsenals and dockyards on the eve of World War 1. Vickers had worked with the Russian Government to establish the Nikolaiev naval shipyard on the Black Sea in 1913 and, at the behest of Canada, had founded a shipyard in Montreal in 1910 to serve the needs of the newly-created Royal Canadian Navy. It was able--as Tom Vickers proudly declaimed in 1909--to build and arm entirely from corporate resources three 'dreadnoughts' within three years.[33] By 1911 the firm was embarking on aircraft manufacture.

Beardmore.
A Glasgow steel-master, William Beardmore started manufacturing Harvey armour plate in 1895. Four years later, he transformed his Parkhead Forge into an armour plate mill and immediately afterwards bought Robert Napier's shipyard at Govan and marine engine works at Lancefield. Beardmore became a shareholder in the Thornycroft destroyer-builder in 1901, regarding it as a potential customer for the guns made at Parkhead (a pipe dream, as it happened). A year later, the firm purchased 90 acres at Dalmuir to lay out a 'state-of-the-art' naval shipyard, engine and boiler works. At the same time, the alliance with Vickers ensured it of a ready market for the Krupp armour processed by Parkhead's rolling mills and huge 12,000-ton forging press. The firm's gun market was guaranteed by its accession to the Admiralty's gun pool which gave it a quota of the Navy's gun orders. A division of labour was instituted within the Vickers-Beardmore combine whereby the former's Barrow Works handled gun mountings of calibres greater than 6-inch while the latter dealt with smaller calibres. Other ventures were not neglected. Beardmores acquired the Arrol-Johnston motor car company in 1903 and, a decade later, had built a new factory at Dumfries for the purposes of designing and constructing cars and aircraft, along with their engines. In addition, it bought and rebuilt the Mossend Steelworks in 1905 to consolidate the supply of plates and sections to the Dalmuir shipyard. Unlike other MIEs, however, Beardmores failed to profit from its armament operations, for the most part because of its inability to recoup the heavy expenditures made on the 'greenfield' Dalmuir site.

159

That works, for example, incurred losses of £25,000
on its first battleship contract and was never able
to fill the available capacity: only ten ships went
down the ways at Dalmuir between 1905 and 1909 in
contrast to the 25 each built by the comparably-
sized (but older and amortised) yards of neighbours
John Brown and Fairfield.[34] Evidently, the connexion
with Vickers offered the company something of a
lifeline at this time.

The machinations of the UK firms were dupli-
cated elsewhere, and integrated arms producers
sprung up in the USA, France and Germany. In a manner reminiscent of the moves pursued by the Sheffield steel firms, the enterprises of Bethlehem,
Schneider and Krupp all transformed themselves into
general defence contractors. As we have stated,
Bethlehem's rise to prominence was inextricably
linked to the favours showered on it by the US Navy.
It was able to adopt the latest in heavy steel forgings technology in part because US Navy Intelligence
passed on these advances to the firm.[35] Joseph Whar-
ton, the firm's founder, actually travelled to
Europe with the blessing of the US Navy in order to
purchase an entire armour-making plant from England
which included the introduction into the USA of a
hydraulic forging press, a much superior apparatus
to the steam hammer used hitherto. From 1894 Bethlehem began exporting armour (in the first instance
to Russia) at US Navy connivance since it was
believed that overseas demand would substitute for
lulls in domestic ordering. Defence sales permitted
the firm to double its profits in 1913-4 when US
steelmaking as a whole experienced a fall in earnings from $119 million to $82 million.[36] Equally
reliant on its favoured status in official quarters,
Schneider & Cie emerged as the linchpin of France's
late-19th century defence industry. Tracing its origins back to the first systematic exploitation of
iron ore and coal at Le Creusot in 1769, Schneider
was founded in 1836 on the basis of a buy-out of the
furnaces, smelters and glassworks already there.
Indispensable to Napoleon's forces in the wars of
the early-1800s, Le Creusot's assets were soon
returned to the building of cannon for the French
army and the firm subsequently acquired a monopoly
in French gun construction. As intimated earlier,
Schneider also became immersed in armour development
and production and, utilising Nasmyth's power hammer
technology at the Chalon-sur-Saône factory after
1841, the firm built up an enviable reputation as a
locomotive manufacturer and as the main supplier of
marine engines to the French navy. Pursuant to its

armour-making ambitions, it had installed in 1876 a
hammer which, at 100 tonnes, was the most powerful
in the country.[37] Just prior to World War 1,
Schneider was actively engaged in reconstructing the
Putiloff Works in St Petersburg, then the Tsarist
Government's principal arms plant. For its part,
Krupp followed a path mapped out by Alfred (1812-87)
and his son Friedrich Alfred (1854-1902) which progressively put the firm among the leaders of gun and
armour manufacturing by the end of the century. The
former, known as the Cannon King, began casting
steel cannon in 1847 and by the 1860s he had developed a vertically-integrated coal to steel organisation. After 1906 the firm came under the managerial
sway of Dr Gustav von Bohlen und Halbech (hence its
restyling as Krupp von Bohlen) and boomed in the
phase of rearmamental instability prior to the World
War. As with Bethlehem, Krupp had needed the backing
of naval circles to enter into significant defence
production (the army sticking by its own arsenal
system). In 1868 he became a navy contractor on the
strength of his massive 500-ton hammer and its use-
fulness in forging armour, guns, and marine engine
shafting. At the time of Alfred's death, the Essen
gun works had made 24,576 cannon: by 1911 the number
had risen to 53,000, of which no less than 27,000
had been exported to 52 countries.[38] Reminiscent of
Armstrong and Vickers, Krupp took over, at the
behest of the government, the Germaniawerft shipyard
in Kiel so as to prepare it as a major battleship
building facility. By the outset of the war, this
yard had managed to build three 'dreadnoughts' in
comparison with the two of Wilhelmshaven Navy Yard
(state-owned), one of Kiel Navy Yard (state-owned),
three of AG Weser (Bremen), two of Vulcan (Stettin
and Hamburg), three of Howaldtswerke (Kiel), four of
Schichau (Danzig) and the five battlecruisers built
by Blohm + Voss (Hamburg).

THE APOGEE AND AFTER

The climax in the fortunes of the traditional MIE
occurred in World War 1, with a recurrent echo of
the climax in the second global conflict twenty
years later. In the intervening period the MIEs were
forced, willy-nilly, to contemplate conversion to
civil industry and the experience was far from being
satisfactory. It is fair to say that rearmamental
instability in the 1930s occurred at the opportune
time for most MIEs, that is, those that had succeeded in surviving as viable MIEs or, indeed,

surviving at all! While the nature of defence items precluded true mass-production in all but a narrow range of product lines, the traditional MIEs had assumed the status of mature firms by the onset of World War 1: a conflict which was to effectively transform some weapons lines into genuine mass-production articles, spur the rapid diffusion of defence production technology to a host of imitator firms both in the AICs and elsewhere, and provoke the founding of novel weapons systems requiring new organisational solutions. Thus a brief waxing of flourishing fortunes was replaced by a discordant era (inclusive of popular accusations of war profiteering) of hesitant demand, and to compound the problems yet further, by the emergence of modern MIEs competing for straitened defence budgets. These trends are illustrated through examples drawn from the experiences of the leading traditional MIEs.

World War 1

The oldest MIE of all, the state arsenal system, was an immediate beneficiary of wartime equilibrium. For example, in the UK the three ROFs (Woolwich, Enfield and the propellant plant at Waltham Abbey) were eventually joined by about 250 newly-created state-owned ordnance establishments. The Ministry of Munitions operated, either directly or in conjunction with private firms in the 'agency' manner, some 40-plus National Shell Factories and 15 National Projectile Factories, respectively for small and large-calibre shells, not to speak of dozens of explosives plants including 19 National Filling Factories, four National Cartridge Factories and three Trench Warfare Filling Factories. One of its most impressive ventures was the construction of an entire community, Elizabethville, in County Durham to accommodate 4,000 Belgian refugee workers employed in its Birtley factory.[39] The arsenal system was also strong in the area of small arms. Enfield supplied about two million rifles, leaving the private MIEs (especially BSA and the London Small Arms Company) to make the remaining 1.75 million. Machine guns, however, remained the preserve of private industry: Vickers and BSA being joined by a subsidiary of Société Hotchkiss, set up in Coventry. Attempts to operationalise a National Machine Gun Factory at Burton-on-Trent were foiled, and this state arsenal was relegated to repair work. In addition, the state ran the Royal Aircraft Factory at Farnborough and was in the process of setting up four National Aircraft Factories and four National Shipyards at war's end. On its own authority, the Admiralty ran a torpedo factory at Greenock and a cordite factory at Holton Heath. However, while the Royal Dockyards vastly expanded in scope and size (including the opening of Rosyth), they were confined by and large to repair and fleet support functions rather than new construction. Accordingly, they produced only 49 of the 1,040 naval ships built in the UK during 1914-18 (albeit six of these were battleships and a further 19 were submarines).[40] The fact remains, though, that belligerents everywhere felt it expedient to boost arms production through state facilities. The USA even went so far as to establish a navy armaments works at Charleston, West Virginia, to make armour plate, gun forgings and projectiles (it was, however, allowed to fall into disrepair immediately after the war). Its Navy Yards were called upon to belatedly supplement the private MIEs in warship construction. The New York Navy Yard laid down its first 'dreadnought' in 1911 and went on to complete another three by 1920. In 1916 the Mare Island Navy Yard near San Francisco was also given a battleship contract. The state role should not be overstated, however. Over the same period, the private MIEs turned out nine battleships, with Newport News being responsible for five of them, the New York Shipbuilding Company of Camden, New Jersey, adding another three, and Bethlehem's Fore River yard completing the total with a single ship.

While the Royal Arsenal (which became the largest munitions plant in the world) spearheaded the UK state's role in artillery production, the mainstay of gun-making—as with warships—was the private MIE. Armstrong alone built one-third of the country's output of guns and mountings (i.e. it produced 13,000) and for good measure, also made 14.5 million shells, 18.5 million fuzes, 21 million cartridge cases, 100 tanks, 1,075 aeroplanes and three airships. Vickers was equally productive; its relatively minor Canadian Vickers subsidiary, for example, turned out 1.5 million shells not to mention 24 submarines and 214 patrol craft. Another of its associates, Beardmore, turned over its Arrol-Johnston Paisley factory to shell making, set up aeroengine plants at Tongland and Coatbridge, erected an airship factory at Inchinnan, and converted its Dumfries factory and part of the Dalmuir works to aircraft manufacture. For its part, Vickers' Barrow shipyard completed seven cruisers and 61 submarines during the war years, as well as a sizeable tonnage of auxiliary vessels. Barrow was equally instrumental in establishing an airship works and shared with Beardmore and Armstrong Whitworth (Selby) in the construction of Admiralty airships. To augment this branch of production even further, the company opted to build a large airship factory and accompanying village at Flookburgh, ten miles from Barrow, but the venture was abandoned owing to severe steel shortages after £792,000 had been expended on it.[41] The company, nevertheless, ended the war as a major aeroplane maker with a large establishment at Weybridge. Not surprisingly, then, four of the traditional MIEs ranked in the top 50 UK companies in 1919. Vickers was in the forefront of them (ranked fourth overall), with an estimated market value of £19.5 million; Armstrong Whitworth came next (eleventh of the top 50) at £12.2 million, while John Brown (eighteenth) and Cammell Laird (thirty-seventh) registered market values of £7.7 million and £4.8 million respectively.[42] Even Palmer had experienced a good war. It soon added shell-making to its shipbuilding efforts and, by the end of the 1914-15 financial year, was able to turn a 1912 loss of £128,413 into a credit of £42,772.[43] In the USA, Bethlehem was charged with operationalising the government's great plan of 1917 to vastly increase shipbuilding capacity. The resultant Emergency Fleet Corporation, a joint government and private enterprise initiative, set up three new shipyards (including the huge Hog Island yard near Philadelphia) and, under Bethlehem's auspices, pioneered the use of prefabricated process techniques in shipbuilding. The scene was set for Bethlehem becoming a vertically-integrated shipbuilder with a yard located adjacent to the steelmaking complex at Sparrows Point, Maryland.

As intimated, the war was instrumental in enticing a host of new entries into the market previously dominated by the traditional MIEs. The majority of the newcomers were 'hostilities only', but the tenure of others became so drawn out that they were conceived as permanent additions to the stock of MIEs. To illustrate this latter phenomenon, one need only point to Swan Hunter. This, the Tyne's largest merchant shipbuilder, had been tempted into Admiralty work in 1908 with destroyer contracts. In the war it was to greatly extend its warship work and thenceforth was never to retreat from defence production. One monitor, two cruisers, 27 destroyers and four submarines left its slipways between 1914 and 1918. Of the numerous merchant yards pressed into temporary duty as naval

suppliers, Harland & Wolff was--like Swan Hunter--
smitten with the desire to join the ranks of permanent MIEs. The firm's Belfast and Govan shipyards
were transformed, in part, into warship facilities:
producing in consequence a light battlecruiser, 13
monitors, a cruiser and six destroyers during the
war years. What is more, the firm also found the
resources to engage in gun production (accomplished,
at the end of the war, through takeover of the Scot-
stoun, Glasgow, factory of the Coventry Ordnance
Works) and aircraft manufacture (undertaken, with
the adaptation of the shipyard joiners plant). In
Germany, meanwhile, the general engineering firm of
Rheinmetall rose from being a relatively minor manufacturer of pistols and machine guns (under the
Dreyse hallmark) to becoming an artillery producer
of major proportions and a permanent fixture of the
country's defence-industrial base. In Italy, Fiat
entered munitions production, making amongst other
things, Revelli machine guns, submarines and tanks.
It resolved to remain in defence production after
the war, setting up an armaments division (SAFAT) to
specialise in machine guns (it was, in fact, sold to
Breda in 1930 although Fiat persevered with defence
involvement engaging, in a major way, in aircraft,
aeroengine and torpedo work). In a similar fashion,
Breda of Breschia was a locomotive maker that converted to machine-gun manufacture during the war,
eventually transforming itself into a significant
arms organisation in the postwar epoch. As a final
example of new entries, Henry Ford's automobile company stands as an insight into things to come;
namely, enterprises stemming from newer technologies
than those underpinning the traditional MIEs but
which, none the less, aspired to compete in traditional defence lines. His was a praiseworthy (if
problematic) attempt to apply assembly-line methods
to the construction of small warships. Indeed, a
Michigan car factory was extended to accommodate the
plant needed to turn out one 'Eagle Boat' submarine-
chaser per day (in reality, one boat per month was
the eventual achievement).[44] While the Ford shipbuilding venture proved to be something less than
successful and was never repeated, the firm was to
become a major, if intermittent, player in defence
production from this time onwards.

The Italian new entries served notice of the
diffusion of defence technology outside of the original core of firms domiciled in the UK, France, Germany and America. We have already noted how the traditional MIEs, consistent with the defence industry
variant of industrial decentralisation that accords

with maturity, transferred aspects of defence production to other countries by way of subsidiaries
and joint ventures with foreign governments. Italy
was an early recipient of those multinational ventures. The main state arsenal at Pietrasa, founded
in 1861, had been joined by the Genoa-based Ansaldo
firm, whose founder was an Englishman. The latter's
interests in shipbuilding and marine and locomotive
engineering had taken a distinct turn towards
defence when it joined with Armstrong after 1904.
The British MIE already enjoyed extensive Italian
interests, having operated the Pozzuoli works in
Naples since 1885. Vickers, Schneider and the German
Ehrhardt concern all had Italian ordnance interests
at either La Spezia or Milan by the beginning of the
war.[45] The Vickers venture, a joint undertaking with
Orlando-Odero (later OTO), built the large Vickers-
Terni arsenal at La Spezia between 1906 and 1910, a
facility which supplied the Italian navy with gun
mountings and armour plate while Ansaldo produced
'dreadnought' hulls. In Spain the British MIEs were
equally active. SECN, the naval dockyard organisation, was rebuilt with their help so as to enable
the country to undertake modern warship construction. It became the largest shipbuilding enterprise
in Spain and was forthcoming with munificent divi-
dends from 1908 until the 1930s. As well as the
yards at Ferrol, Cartagena, Sestao (Bilbao) and San-
tander, SECN created the Carraca (Cadiz) and Reinosa
heavy gun, gun mounting and shell plants, together
with the Ferrol and Cartagena turbine engine shops.
Russia, too, was not averse to using imported technology. With the assistance of Schneider and St Cha-
mond from France, the armaments assets at Bryansk,
the Putiloff Works and the Nevsky Steel Casting &
Forging Works were made operational. With the aid of
John Brown, Armstrong Whitworth and Vickers, the
Nikolaiev dockyard and Tsaritsyn (Volgograd) arsenal
were put in order.

Russia's late enemy, Japan, was not slow to
call on the MIEs either. The Meiji had restored cannon factories at Tokyo and Osaka. The former concentrated on small arms, starting in 1870 with an
Enfield design, while the latter was dedicated to
artillery, beginning with French four-pounders in
1872. In that same year, the first naval arsenal was
created at Yokosuka, to be followed by facilities at
Kure/Onohama (founded by an Englishman, Kirby) in
1884, Sasebo in 1889 and Maizuru in 1901. By the
early years of the century, Sasebo, Kure and Maizuru
were undertaking warship construction and were
backed up by a naval gunpowder factory at Tokyo and

a torpedo works at Ominato. The Imperial Navy also owned marine engine works at all those places and Bako and Ryojun besides. Significantly, however, Japan continued to import Curtis turbines from Bethlehem in the USA and Parsons turbines from Newcastle (England), not to speak of Elswick pattern guns, until the outbreak of war. A private shipbuilding industry had been allowed in 1884: the two most prominent firms entering it being Mitsubishi at Nagasaki and Kawasaki at Kobe.[46] Both began 'dreadnought' building in 1912 and, over the next decade, completed two apiece to supplement the three built at the Yokosuka Navy Yard and the two built at the Kure Navy Yard. Yet, despite Japan's attempts to create state MIEs and encourage private firms to join them in shouldering the arms expansion burden, its defence industry could not escape the influence of the chief European and American MIEs. In addition to building the designs of foreigners, the Japanese were compelled to call upon their expertise (Vickers and Armstrong) in order to erect armour-making facilities and, in consequence, were obliged to give them a stake in the resultant enterprise.

Postwar Doldrums

The comprehensive cancellation of defence contracts at the Armistice came as a rude shock to the MIEs and the subsequent period of disarmamental instability and peacetime equilibrium was to rock them to their foundations. Some firms had attempted to develop alternative civil products during prewar hiccups in defence contracting: the shipbuilders opted for merchant vessels while small-arms firms such as BSA and FN became significant cycle manufacturers, and these ploys were tried again.[47] Anticipating a boom in replacement merchant shipping, Palmer bought the Amble shipyard and a ship-repair facility at Swansea, Beardmore added three merchant shipbuilding berths to its Dalmuir yard, and both Browns and Fairfield re-equipped their war-built West Yards to produce tankers. Other defence capacity was less easily diverted to civil uses. Only three of the wartime state ordnance factories survived in working order: the filling factories at Hereford and Birtley and the explosives plant at Irvine. Woolwich itself was drastically cut back, and the part that remained was driven to manufacture railway locomotives in order to retain a cadre of skilled workers. Locomotive manufacture was grasped by the private MIEs as well. Armstrong Whitworth,

Table 4.3 : Vickers and Demobilisation, 1919

Works/Facility	Defence Speciality	Postwar Conversion
Barrow	warships, guns	merchant ships, locomotives, boilers turbines, diesels
Crayford	machine guns	sporting guns, sewing machines, car parts
Erith	machine guns, shells	matchmaking machinery, machine tools, cardboard & box-making machinery, gas meters
Birmingham	cartridge cases	mass-produced cars
Dartford	ammunition	furniture, wooden toys, washing machines
Eskmeals	heavy gun range	-
Sheffield	armour	rails, forgings, stampings

Source: J.D. Scott, Vickers: A History, (Weidenfeld & Nicolson, London, 1962), p. 48 and 140.

Vickers, Hawthorn Leslie and Beardmore all converted their gun or warship engine plants to locomotive building. Indeed, the general defence contractors systematically went about boosting their railway offerings. Vickers bought the Metropolitan Carriage concern, John Brown acquired the Craven Railway Carriage & Wagon Company, Cammell Laird took over the Midland Railway Carriage & Wagon Company, while the Coventry Ordnance Works disappeared altogether, emerging in disguised form as the English Electric Company with constituents that included the Preston locomotive-building firm of Dick Kerr & Company. As an alternative, the automotive industry frequently found favour. Beardmore, for example, converted its Paisley shell plant to taxi-cab construction, its Anniesland fuze factory to heavy car assembly, bought the Glasgow van-maker J. H. Kelly, and established Beardmore Motors at Kings Norton for commercial vehicle manufacture.

Vickers also bolstered its automotive interests, but much else besides. As Table 4.3 makes clear, it systematically reassigned its defence plants to civil lines encompassing a plethora of

168

Figure 4.5 : Pre-Merger Vickers and Armstrong Groups

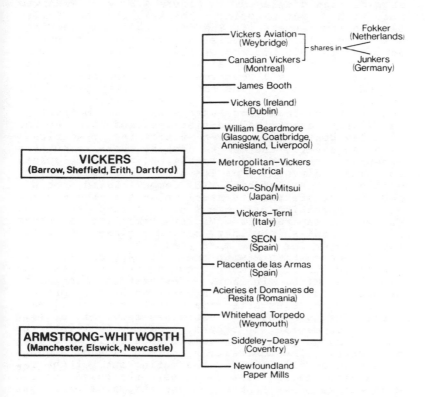

items from gas meters to sewing machines. Armstrong
Whitworth also contemplated a variety of substitute
activities, the most ambitious and bizarre being the
Newfoundland Paper Mills scheme which committed the
firm to the development of a new port and paper mill
at Cornerbrook and an expenditure of £3 million.
Gross mismanagement eventually drove costs up to
over £11 million and the parent firm was tottering
on the brink of collapse. This fact, combined with
Vickers' inability to make a success of its conversion programmes, led to the two firms merging in
1927 at government instigation. For the sake of survival, the resultant Vickers-Armstrong concern drastically pruned its interests. Given the unpromising

state of defence demand, armaments establishments
were ripe for rationalisation. The multiplicity of
subsidiaries displayed in Figure 4.5 were winnowed
out, with, for example, the sale of the fully-
fledged arms assets of Beardmore and Canadian Vickers along with the specialist firm of James Booth,
acquired in 1915 because of its manufacture of dura-
lumin materials used in airships (Table 4.4). Neither were the civil ventures spared. The brave hopes
pinned on consumer items had been quickly overturned
and along with them went Vickers-Armstrong's interest in the Newfoundland Paper Mills, the Metropoli-
tan-Vickers electrical initiatives, and Wolseley and
Siddeley-Deasy cars. Even within the traditional
defence context, the merged company chose to cut the
extent of its vertical integration. In particular,
that meant divestment of its steel interests. Along
with another retrenching MIE, Cammell Laird, the new
combine transferred its steel plants to the English
Steel Corporation (Figure 4.6). A rump Armstrong
Whitworth, separate from Vickers-Armstrong, survived
as a merchant shipbuilder, marine engineer and railway engineer for a few more years.[48]

Disenchantment with the efficacy of economic
conversion to civil manufacturing became widespread
among the MIEs. In fact, few of them managed to
bridge the divide between success as defence producers and success as effective operators in civil markets. The outcome was restrained survival for some
firms, but demise for others. Palmer fell into the
latter category while Beardmore escaped it merely by
a whisker. The Admiralty's introduction of a quota
system which allowed for the meting out of the few
replacement naval vessels across the board to the
MIEs did not save Palmer's Tyne shipyards or Beardmore's Dalmuir shipyard. In a relative sense, the
Royal Dockyards prospered most by it, usurping the
private MIEs in cruiser and sloop supply.[49] The private firms, however, benefited greatly from the fact
that they dominated the supply of 'systems' going
into the hulls built at state yards. For example,
Beardmore, Hawthorn Leslie, Parsons, Fairfield, Cammell Laird and Vickers-Armstrong supplied all the
propelling machinery used in Royal Dockyard-built
cruisers of the 1920s. By the same token, Vickers'
Barrow works earned profits in 1928 of more than
£250,000 on sales of heavy guns and mountings in
marked contrast to the nominal £15,000 derived from
shipbuilding.[50] In lieu of anything else, Vickers-
Armstrong decided to concentrate on those defence
offerings deemed to have market possibilities; thus,
in 1928, it acquired the Supermarine firm in order

Table 4.4 : Formation of Vickers-Armstrong MIE

Rationalisation	Assets
sale of Newfoundland Paper Mills	Barrow shipyard
sale of Metropolitan-Vickers	Walker Naval Yard (closed 1928-30, 1931-4)
sale of Wolseley	Weybridge aircraft plant
sale of interest in Beardmore	Barrow and Elswick gun and mountings plants
sale of James Booth & Co	Erith and Crayford fire control, machine gun, gun carriage, small naval guns and mountings plants (Erith closed 1931)
sale of Canadian Vickers	Established tank plant at Elswick (1928)
merger of steel interests into English Steel Co	Acquired Supermarine Co aircraft plant at Southampton (1928)
merger of rolling-stock interests into Metropolitan-Cammell	

Source: J.D. Scott, Vickers: A History, pp. 167-89.

to strengthen its military aircraft business and set up a tank plant at Elswick with the purchase of patents for Carden-Loyd tanks and armoured cars (another shipbuilder, Harland & Wolff, also flirted with tank production a little later).   Elsewhere in the defence field the company continued to rationalise: the Walker warship yard was closed in 1928-30 and again in 1931-4 while Erith was expunged in 1931 on the transfer of its ordnance equipment to Crayford. All the firm's Spanish, Japanese and Italian defence ties were excised in the early-1930s too. Other traditional MIEs attempted to follow suit. Schneider entered the submarine field in 1922 and, in the general defence market, attempted horizontal integration across Europe through its replacement of Krupp as the controlling agent of the great Skoda works in Czechoslovakia. As a result, Schneider-Creusot gained access to East European and Asian markets which the Pilsen gun, munitions and tank producer had seriously penetrated with the enthusiastic backing of the Czech state. Other initiatives, arranged through the Union Européenne Banque connection, gave the French MIE holdings in the aircraft

Figure 4.6 : MIE Divestment of Steel Interests

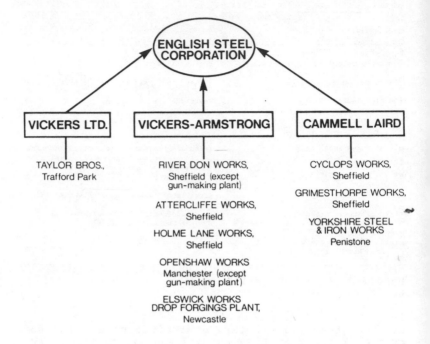

and munitions plants at Brno, the gas and explosives factories at Marienberg, Asee and Olomouc, and the aeroengine plant at Warsaw. Reputedly, it also bought into the Swedish Bofors gun concern in reaction to an attempt engineered by Krupp in 1927 to gain control of that firm.[51]

Diversification into military aviation was another sign of the times. In the 1930s Vickers was joined by other shipbuilders keen on gaining access to a new field of military technology. Among the more memorable attempts at entering aircraft manufacture were those of Harland & Wolff (as Short & Harland), Swan Hunter (via Airspeed), Blohm + Voss (through its 1933 formation of Hamburger Flugzeugbau) and Cantieri Riuniti dell'Adriatico (which set up in 1931 a flying boat factory alongside its shipyard at Monfalcone). Earlier, Japanese shipbuilders Mitsubishi and Kawasaki had entered aircraft manufacture when, in 1919, they had accompanied the navy arsenals at Hiro and Yokosuka into this promising

new area of defence production. The stirring in
defence activity which occasioned the related diversifications into aircraft was to swell into a
Europe-wide rearmamental movement as the 1930s
unfolded, mirrored in the Far East through Japanese
belligerency. Pickings once again were in the offing
for those MIEs remaining in business.

A government preference for supporting its own
Navy Yards had nothing short of catastrophic effects
on US private MIEs. Newport News got through the
1920s by dint of temporary diversification into rail
cars, traffic lights, hydraulic turbines and structural steel. Submarine specialist Electric Boat had
no US submarine contract from 1918 until 1931. It
survived on the income earned from building ferries,
tugs and the bridges for Connecticut's Merritt Park-
way. Bethlehem remained viable on account of the
range of its diversified interests, although paring
of defence capacity was widespread. The Cramp Ship &
Engine Company of Philadelphia and the Bath Iron
Works did not survive at all, diversification into
merchant shipbuilding being insufficient to carry
them over the trough in naval ordering (although
both were revived on the strength of rearmamental
orders prior to World War 2). Orders to private
builders of submarines were only reinstated in 1932:
too late to save the Bliss enterprise and the Lake
Submarine Company of Bridgeport, Connecticut.[52] As
in Europe, it was rearmament which saved American
defence industry, and as that defence expansion was
couched largely in terms of combating Japan, it
offered considerable scope to the traditional MIEs
and their preoccupation with warships. By the end of
the 1930s the three armour-plate manufacturers had
been resuscitated and the US Navy had relented in
its policy towards private shipbuilders. Indeed,
private firms were given all design and construction
jobs other than for battleships, aircraft carriers
and submarines, and even with those classes private
enterprise was drawn into production. Moreover, new
firms joined the longstanding firms in this revival:
GE and Westinghouse, for example, came to virtually
monopolise orders for marine engines. Yet, everywhere, the old masters were busily engaged in building a new generation of 'dreadnoughts' on the eve of
World War 2. In Britain the task was shared by Vickers-Armstrong, Cammell Laird, John Brown, Fair-
field--and that successful interloper--Swan Hunter.
In the USA the Navy Yards were supplemented by Newport News, the New York Shipbuilding Company and
Bethlehem's Quincy (ex-Fore River) yard. France, as
usual, called on the Arsenal de Brest, but also

173

pressed the Penhoët yard at St Nazaire into service. Hitler's new capital ships were allocated to Wilhelmshaven Navy Yard, Blohm + Voss and Deutsche Werke, Kiel (the HDW predecessor), while Japan turned to Mitsubishi and the Navy Yards at Kure and Yokosuka.

Renaissance

The dramatic revival of the traditional MIEs in connexion with World War 2 is nowhere more vividly illustrated than in Germany. Despite wholesale abandonment of arms capacity after the earlier global conflict, German defence expertise had remained alive--if somewhat scattered--through the employment of German designers and engineers in foreign facilities. One of the most interesting repercussions of the foreign involvements forced on German engineers was the creation, in 1923, of the Werkzeugmaschinenfabrik Oerlikon in Switzerland. Oerlikon was a moribund machine-tool maker, put on its feet by the German Magdeburg engineering concern. Under the tutelage of Emil Bührle (who orchestrated a management buy-out in 1937), the firm acquired the rights to the Semag-Becker cannon and had become a major exporter of automatic cannon by 1939. Similarly, the Bofors company in Sweden benefited from strong German influences (not least that of Krupp), and between them, Oerlikon and Bofors came to dominate the scene in light anti-aircraft artillery in World War 2.[53] With Hitler's rearmament drive back home, however, the skills of the expatriates reverted to the task of operationalising capacity in the Reich. For example, submarine construction was revived in 1935 in three yards; namely, Krupp's Germaniawerft, the state-owned Deutsche Werke and the Deschimag concern of Bremen. Yet, by 1939, no fewer than 13 extra yards were incorporated into the submarine programme, and with the outbreak of hostilities, a further three were brought on stream. During the decade of Hitler's programme, those yards produced 1,158 U-boats, of which an impressive total of 222 stemmed from the efforts of Blohm + Voss, 165 came from Deschimag, 128 from Germaniawerft, 114 from Deutsche Werke and 90 from Schichau (Danzig).[54] Krupp again came to the fore in across-the-board arms production, although it was rivalled by relative newcomers such as the Rheinmetall-Borsig concern which ran twelve major arms factories. In a strange twist of fate, the resources of the Skoda Works in Czechoslovakia became, like

174

Krupp, a cornerstone of the Reich's heavy ordnance industry.

As for the UK, revival pervaded all traditional MIEs from the ROFs to the relics of the vertically-integrated firms. In reference to the former, the government had instituted a substantial expansion scheme just prior to the war and, in so doing, had taken cognizance of the need to disperse the factories to sites remote from German airfields (so long as the sites were not handicapped in terms of labour supply). By the beginning of the war, new filling factories had been erected at Chorley in Lancashire and Bridgend and Glascoed in Wales. These were soon augmented by filling factories located, in the main, in the North of England (Swynnerton, Risley, Kirby, Thorpe Arch and Aycliffe). Large expanses of land in combination with large pools of labour were judged vital for propellant plants and they appeared at such places as Bishopton (Scotland) and Wrexham (North Wales). Remoter sites were preferred for explosives plants, as the ROFs at Pembrey (South-West Wales), Drigg and Sellafield (Cumbria), and Bridgwater (Somerset) duly attest. A new category of ordnance plant, the Engineering ROF, was adapted to the needs of mechanised armies. The breakdown of the sites of these ROFs by region gives some indication of the importance attached to geographical dispersion: North-West England (eight), North-East England (five), South-West England (three), the Midlands (two), the Home Counties (two) and the South East (one).[55]

As far as the private MIEs are concerned, the lapsed vertical linkages were informally brought to bear again. In consequence, Vickers-Armstrong, Beardmore, Harland & Wolff, Firths and the English Steel Corporation became fully-occupied gunsmiths—and much else besides. As well as the Barrow and Walker shipyards, Vickers-Armstrong ran gun and tank plants at Elswick and Scotswood (Newcastle), main aircraft factories at Weybridge, Eastleigh (South-ampton) and South Marston (Swindon), and 'agency' (shadow) aircraft factories at Castle Bromwich, Chester and Blackpool. Its shipyards produced a battleship, eight aircraft carriers, six cruisers and monitors, 35 destroyers and the bulk of the UK's tonnage of submarines. As Table 4.5 clearly shows, this concerted effort across the spectrum of warship types was shared with Cammell Laird, John Brown, Swan Hunter and Fairfield (although the last three kept out of submarine work). Harland & Wolff served as a specialist supplier of aircraft carriers, escorts and minesweepers, while destroyers were

Table 4.5 : Warship Output of British MIEs, 1939-45

Firm	Battleships	Aircraft Carriers	Cruisers and Monitors	Destroyers/ Minelayers	Submarines
Vickers-Armstrong	1	8	6	35	106
Cammell Laird	1	1	3	26	39
John Brown	2	1	4	26	-
Swan Hunter	1	2	4	30	-
Fairfield	1	2	3	21	-
Harland & Wolff	-	6	2	-	-
Stephen	-	1	3	16	-
Hawthorn Leslie	-	1	3	21	-
Devonport Dockyard	-	1	1	-	4
Scotts	-	-	4	18	17
Chatham Dockyard	-	-	1	-	11
Portsmouth Dockyard	-	-	1	-	2
White	-	-	-	26	-
Thornycroft	-	-	-	22	-
Denny	-	-	-	11	-
Yarrow	-	-	-	20	-

handed to the specialists in that trade: White, Thornycroft, Yarrow and--since the earlier world war--Denny.[56] The Royal Dockyards were largely given over to repairs, but they managed to retain some new construction capabilities, especially in the submarine field. Firms such as Alex Stephen (Linthouse, Glasgow), Scotts and Hawthorn Leslie fell midway between the large producers and the smaller specialist producers, keeping in tune with both by developing capabilities that stretched across several warship types. These warship constructors were supported by a veritable host of merchant shipyards pressed into the building of smaller naval vessels and standardised cargo ships, not to speak of the nine Royal Naval Factories operated on an 'agency' basis.

New entries were equally prominent in the USA. Most impressive by far were the new facilities that arose out of the efforts of industrialist Henry Kaiser, who almost single-handedly, became the driving force in the US shipbuilding industry. Not only was he instrumental in founding several emergency shipyards--California Shipbuilding, Kaiser Vancouver,

Table 4.6 : Producers of Liberty Ships

Yard	Building Berths	Management	Resultant organisation	Liberty ship output
Mobile, ALA	4	Alabama Drydock & Shipbuilding Co (existing shipbuilder)	Alabama Drydock Co, Pinto Island	20
Baltimore, MD	16	Bethlehem Steel Co (existing shipbuilder)	Bethlehem-Fairfield	385
Los Angeles, CA	14	Henry Kaiser	California Shipbuilding Corp, Terminal Island	336
New Orleans, LA	86	American Shipbuilding Co (existing shipbuilder)	Delta Shipbuilding Co	188
Brunswick, GA	6	J.A. Jones Construction Co	Jones Brunswick Yard	85
Panama City, FLA	6	J.A. Jones Construction Co	Jones Wainwright Yard	102
Vancouver, WASH	12	Henry Kaiser	Kaiser Vancouver Yard	10
Sausalito, CA	6	Kaiser Group (W.A. Bechtel)	Marinship Corp	15
South Portland, ME	6 & docks	Bath Iron Works, then Todd Shipyards Corp (existing shipbuilders)	New England Shipbuilding Corp	244
Wilmington, NC	9	Newport News Shipbuilding Co (existing shipbuilder)	North Carolina Shipbuilding Co	126
Portland, ORE	13	Henry Kaiser	Oregon Ship Building Corp	322
Richmond, CA	19	Henry Kaiser	Permanente Metals Corp, Yards No 1 and 2	489
Jacksonville, FLA	?	local ship-repair firm + New York contractors	St. Johns River Shipbuilding Co	82
Savannah, GA	6	local construction firm	Southeastern Shipbuilding Corp	88
Houston, TX	9	Todd Shipyards Corp (existing shipbuilder)	Todd Houston Shipbuilding Corp	208
Providence, RI	6	Kaiser Group + Walsh Construction Co	Walsh-Kaiser Co	11

Source: Compiled from L.A. Sawyer and W.H. Mitchell, The Liberty Ships, 2nd edn (Lloyd's of London Press, Colchester, 1985).

Marinship, Oregon Ship Building, Permanente Metals and Walsh-Kaiser (see Table 4.6)--but he was responsible for the introduction of the steel industry to California. His Fontana integrated mill annually produced 675,000 tons of ingot steel for the purpose of supplying the West Coast shipyards. All told, the Kaiser yards produced record tonnages of standard cargo ships (Liberty and Victory types) and large numbers of escort carriers (small aircraft carriers built on cargo ship hulls), naval auxiliaries and landing craft (and Kaiser maintained his shipbuilding interests into the postwar era courtesy of the San Diego shipyard of the National Steel & Shipbuilding Company). As Table 4.6 evinces, the traditional MIEs were also drawn into the emergency shipbuilding scheme. Bethlehem, Bath Iron Works and Newport News all ran new yards, as did the established merchant shipbuilders of Alabama Drydock, American Shipbuilding and, above all, the Todd group. Within a remarkably short period, formerly inexperienced naval builders came to fully complement the traditional MIEs in warship construction. The large and complex ships--battleships and aircraft carriers--were reserved for the clutch of Navy Yards (New York, Philadelphia and Norfolk) and traditional private constructors (Bethlehem, New York Shipbuilding and Newport News). Bethlehem was also heavily committed to cruiser and destroyer production, and it arranged its yards accordingly: the capital ships being undertaken by Quincy, the cruisers shared between Quincy and San Francisco, the destroyers emanating from those two facilities and yards at Staten Island and San Pedro, while the destroyer escorts were assigned to Hingham, San Francisco and Quincy. Firms with defence connexions that had been allowed to wither; namely, Federal (of Kearny and Newark) and Cramp, were co-opted into the building of cruisers and smaller warships. Table 4.7 summarises the output of warships, indicating that a vast tonnage of destroyers derived from new-entry producers (most notably, Consolidated of Orange, Texas). The minor Navy Yards were also incorporated into this effort and, reminiscent of the UK experience, one such state yard (Portsmouth) matched the private MIE prime mover (Electric Boat) in the specialist area of submarine construction.

A final insight into the performance of traditional MIEs at this time can be had from an overview of the Japanese and Italian warship concerns. Both latecomers, their arms, armour and shipbuilding enterprises had gained from technology transferred from the pioneers: to such an extent, indeed, that

Table 4.7 : US Warship Output, 1941-5

Firm	Battleships and battlecruisers	Aircraft carriers[1]	Cruisers	Destroyers and destroyer escorts	Submarines
Bethlehem	1	5	17	205	-
New York SB	4	11	14	-	-
Newport News	1	11	8	-	-
Norfolk Navy Yard	1	3	-	12	-
New York Navy Yard	2	4	-	-	-
Philadelphia Navy Yard	2	3	2	20	-
Federal	-	-	4	131	-
Cramp	-	-	5	-	14
Bath Iron Works	-	-	-	72	-
Boston Navy Yard	-	-	-	86	4
Charleston Navy Yard	-	-	-	32	-
Seattle-Tacoma	-	-	-	31	-
Puget Sound Navy Yard	-	-	-	16	1
Gulf SB	-	-	-	7	-
Consolidated, Orange	-	-	-	152	-
Todd-Pacific, Seattle	-	-	-	12	-
Mare Island Navy Yard	-	-	-	34	18
Dravo	-	-	-	12	-
Defoe	-	-	-	17	-
Brown, Houston	-	-	-	41	-
Western Pipe & Steel	-	-	-	15	-
Tampa SB	-	-	-	12	-
Electric Boat	-	-	-	-	86
Portsmouth Navy Yard	-	-	-	-	75
Manitowoc	-	-	-	-	28

Note: 1. excludes escort carriers built from mercantile hulls.

Table 4.8 : Japanese Warship Output, 1941-5

Firm	Number of ships launched: Aircraft carriers	Cruisers	Destroyers and destroyer escorts	Submarines
Mitsubishi, Nagasaki	3	-	5	-
Mitsubishi, Yokohama	-	1	-	-
Mitsubishi, Kobe	-	-	-	28
Kawasaki, Kobe	3	1	1	28
Kawasaki, Tanagawa	-	-	-	7
Kure Navy Yard	3	1	-	24
Yokosuka Navy Yard	1	1	17	12
Sasebo Navy Yard	-	3	2	28
Maizuru Navy Yard	-	-	21	-
Fujinagata, Osaka	-	-	16	-
Uraga Dock, Tokyo	-	-	8	-
Tama Zosensho, Tamano	-	-	-	6

by the time of World War 2 they had mastered the
traditional technology and in some respects exceeded
the capabilities of their former mentors. Japan's
zaibatsu conglomerates, Mitsubishi and Kawasaki,
kept pace with the state Navy Yards in governing the
development and supply of major fighting ships.
Table 4.8 indicates that a rough division of interest was imposed.   Thus, it is evident that aircraft
carriers were obtained from the two private MIEs and
the two state dockyards of Kure and Yokosuka, cruisers were procured from Mitsubishi and the three
dockyards of Kure, Yokosuka and Sasebo, while contracts for smaller vessels were dispersed more
widely. The Maizuru Navy Yard, Fujinagata and Uraga
yards built destroyers to the exclusion of all else,
whereas the Tama Zosensho (a Mitsui affiliate) confined itself to submarines.   A second division of
labour existed within the conglomerates. Thus, Mitsubishi's Yokohama yard built a cruiser while its
Kobe yard launched 28 submarines, and together they
functioned as satellites to the Nagasaki complex.

For its part, Kawasaki had its Senshu Works at Tana-
gawa build submarines in contrast to the Kobe Works
which handled aircraft carrier and destroyer work as
well.
    Italy, meanwhile, fostered private MIEs. Over
the span of naval expansionism consonant with Musso-
lini's rule, the number of shipbuilding entries was
stepped up while foreign influences on them were
reduced (e.g. the links between Vickers-Armstrong
and Ansaldo and OTO were severed, while the Naples
yard of Pattison was liquidated). Table 4.9 encapsulates the results of this effort. It shows that
Ansaldo and the Trieste yard of Cantieri Riuniti
dell'Adriatico monopolised battleship tonnage, but
they shared cruiser work with several other firms,
the most important of which were the Leghorn and
Muggiano yards of Cantieri Odero Terni Orlando
(OTO). As in other countries, destroyers were dele-
gated to both major and minor shipbuilders (e.g.
OTO's Sestri Ponente yard and Cantieri del Tirrero's
yard at Riva Trigosa). Submarines, however, were
assigned to three specialists: OTO Muggiano, Tosi
(at Taranto) and the Monfalcone yard of Cantieri
Riuniti dell'Adriatico.

SENESCENCE AND SUCCESSORS

The end of the war concentrated the minds of MIE
directors on the acute problems of conversion. In
many respects, a replay of the era ushered in by the
Armistice with its dire consequences for the MIEs
was to follow. The Royal Dockyards, for example,
were temporarily given over to merchant ship-repair
and the welding of piping and stainless steel reac-
tor vessels for the Calder Hall (Sellafield) nuclear
plant.[57] New construction of warships was progressively scaled down in the dockyards, to cease altogether in 1970 after Chatham had produced, postwar,
nine submarines, and Devonport and Portsmouth,
between them, had built eleven frigates. A similar
trend was discernible in the USA, with the government phasing out new construction in the Navy Yards
during the 1960s. Instead, both countries chose to
obtain new warships from private MIEs which were
presumed better able to control costs, especially as
they were clamorous for naval contracts in an environment of reduced peacetime-equilibrium orders
(recall Figures 1.1 and 1.2). A shake-out of traditional MIEs was inevitable in spite of the exclusion
of competition from the dockyards (recollect Chapter
3) and industrial concentration was the logical

Table 4.9 : Italian Naval Shipbuilding, 1922-43

Enterprise	Number of ships launched: Battleships	Cruisers	Destroyers and destroyer escorts	Submarines
Cantieri Riuniti dell' Adriatico:				
Trieste	2	5	8	-
Monfalcone	-	-	-	62
Cantieri Odero Terni Orlando:				
Leghorn	-	6	16	-
Muggiano	-	3	-	39
Sestri Ponente	-	-	8	-
Ansaldo	2	7	16	-
Stabilimento Tecnico Triestino	-	2	-	-
R. Cantieri di Castellammare di Stabia	-	2	-	-
Cantieri Navali Riuniti:				
Ancona	-	2	8	-
Palermo	-	1	4	-
Pattison	-	-	4	-
Cantieri Navali del Quarnaro	-	-	9	2
Cantieri del Tirrero	-	-	14	-
Officine & Cantieri Partenopei	-	-	2	-
Bacini e Scali Napoletani	-	-	2	-
Naval meccanica	-	-	4	-
Tosi	-	-	-	39

outcome. The much reduced set of warship construc-tors came to operate in a context of 'permanent receivership', that is to say, their continued exis-tence increasingly came to depend on follow-on defence business. In turn, governments were alerted to the necessity of virtual quota systems in dis-bursing contracts in order to retain the MIEs in being. Since profits were no longer significant, many enterprises signed over the traditional defence activities to formal state control. In other words, putative senescence in shipbuilding, arms and armour manufacturing was expressed through state acquisition of functions no longer valued by private business. They became, in effect, wards or mendicants of the state. To make matters worse for the AICs, parallel-production sources of traditional weapons platforms was increasingly taken up by the NICs in the post-World War 2 years. Thus, the classic relief of resorting to exports when domestic demand was deficient became more and more difficult for the AIC producers that were faced with new-entry competitors located in the areas which, heretofore, had comprised their chief overseas markets.

Permanent Receivership

Examples of the private MIEs succumbing to permanent receivership are legion and several will be touched on here. Schneider is a case in point. Initially shorn of its ordnance business in the pre-World War 2 nationalisation binge of the Front Populaire, the Schneider group moved, postwar, into nuclear engineering as well as revamping its mechanical engineering and steel interests and retaining a shipbuilding stake through the Chantiers du Nord et de la Méditerranée yards at La Seyne, La Ciotat and Dunkirk (a holding which, alas, was faced with liquidation in mid-1986 as a consequence of the collapse of merchant shipping demand). Reconstituted as Creusot-Loire after a 1970 merger, the firm was effectively nationalised in 1982 following massive losses on its steel operations and was bailed-out, in part, because of its longstanding role as France's principal armour manufacturer. At that juncture, virtually all traditional MIE functions in France were in state hands. Under its sway was the dockyard organisation--DTCN--together with its three facilities for new construction; namely, Brest (focusing on aircraft carriers, frigates and cruisers), Cherbourg (building submarines) and Lorient (for destroyers and frigates); as well as a torpedo factory (l'ECAN de Saint-Tropez), tactical weapons plant (l'ECAN de Ruelle) and marine engine factory (l'ECAN de Indret).[58] Shipbuilders outside of the arsenal system were confined to export work--as witness the Saudi F2000 frigates built in the 1980s at the CNIM La Seyne yard. In the UK, too, the armour interests of the steel industry eventually succumbed, with the industry as a whole, to state ownership, as, for a spell, did all the surviving members of the group of private shipyards indulging in warship work.

UK naval shipbuilders had, by all accounts,

been rationalised to the point in the early-1970s
where they retained just sufficient capacity to meet
the needs of the Royal Navy while maintaining a 30
per cent surplus to accommodate export demand.[59] A
combination of reduced UK ordering and vanishing
overseas markets set the scene for state acquisition
in 1977. Nevertheless, defence-dependence increased
during the episode of state ownership. It was
bruited abroad that the five BS yards engaging in
warship activity saw the proportion of their turnover attributable to MoD orders increase from 57 to
72 per cent between 1979 and 1983. They managed to
make a profit of £185 million over the 1980-4 term
in sharp contrast to the corporation's overall loss
of £525 million.[60] These aggregate figures, of
course, masked oscillations in the performances of
individual yards: Brooke Marine of Lowestoft, for
instance, averaged annual profits of £1 million even
though losses of £1.6 million were made in the final
year and the workforce was cut from 850 to 630.[61]
For £162 million, or £248 million below their book
value, BS sold Yarrow, VSEL (the former Vickers Barrow complex) and Vosper Thornycroft. Other firms
with naval building capabilities—Swan Hunter,
Brooke Marine, Scott-Lithgow and Hall Russell (Aber-
deen Shipbuilders)—were also released to the private sector by BS. Significantly, VSEL had been
enlarged prior to privatisation so as to be able to
incorporate the Birkenhead shipyard. This, the former Cammell Laird yard had been kept going solely on
the strength of two Type 22 frigates awarded to it
by the MoD in January 1985. It is noteworthy that
the state retained control of Harland & Wolff (fully
nationalised in 1975) notwithstanding the fact that
the Belfast enterprise was gravitating towards
defence reliance.[62] Yard purchasers held out hopes
of reaping returns from warship specialisation. GEC
acquired Yarrow on account of its long-term frigate
contracts, Swan Hunter's management buy-out was made
feasible by the prospect of orders for frigates and
naval auxiliaries (and, incidentally, required
ditching of the former Vickers-Armstrong Walker yard
in favour of concentration on Wallsend), while the
rationale for a comparable buy-out of Vosper Thornycroft rested on its frigate, fast attack craft and
minehunter expertise. The largest and most diverse
of the bunch, VSEL, builds both nuclear and diesel-
engined submarines at its Barrow shipyard, has
assigned its Birkenhead shipyard to working on a
£300 million order for three Upholder-class diesel-
engined submarines, and undertakes an array of ordnance activities at the Barrow Engineering Works,

including the manufacture of howitzers, missile launchers and naval guns. Privatisation is the lot earmarked for the surviving Royal Dockyards and their task of ship-repair.[63] In accord with this goal, private firms were asked to bid in 1986 for the job of managing warship repairs and refits at the Devonport and Rosyth dockyards.[64]

The UK state arsenals, functioning under the RO banner, were also set on a course for privatisation. To this end, the government announced on 2 April 1987 that BAe had won the tender to buy RO: indication, if such be needed, of the aerospace firm's desire to become a general defence contractor.[65] The UK is not alone in privatising ordnance production. It was pre-empted by Canada. There, the government sold Canadian Arsenals and its two plants (near Montreal and Quebec City) to an engineering firm in 1986. Amassing a profit of C$11.3 million in 1985, the state shell-maker was sold to the SNC Group for C$92.2 million. For its part, RO made a pre-tax profit in 1985 of £26 million on sales of £487 million. At that time, it operated ROFs at Bishopton and Powfoot in Scotland; Birtley, Leeds, Patricroft, Blackburn, Chorley and Radway Green in the North of England; Glascoed in Wales; Featherstone, Summerfield and Nottingham in the Midlands; Bridgwater in the South-West and Westcott, Enfield and Waltham Abbey in the South-East (the Woolwich arsenal having long since been closed). Three of these facilities were run on an 'agency' basis: Powfoot by Nobel Explosive Company, Featherstone by Wickman Wimet and Summerfield by Imperial Metal Industries. The factory at Leeds was geared to MBT manufacture and was sold to Vickers during 1986 as the initiating round of the privatisation plan. Nottingham, meanwhile, built combat engineer tractors and 105mm artillery while Enfield continued its longstanding preoccupation with small arms, commencing production of the SA80 rifle in 1985 (although it had lost, owing to competitive tendering, much of the cannon market to BMARC, the UK subsidiary of Oerlikon-Bührle and, partly in consequence, is slated for closure in 1988).[66] The establishment at Westcott dealt with rocket motor propellants while the other ROFs were devoted to ammunition, explosives, filling and other sundry tasks.

Arms makers elsewhere in Europe, however, remain firmly wedded to state ownership. GIAT continues to supply the French forces--and a goodly number of export customers--with arms flowing from Tarbes, Bourges, Rennes, Le Mans, Tulle and Issy-les-Moulineaux (not to speak of the AFV plants).

Creusot-Loire is complementary, building armoured hulls for tanks and forgings for their weapons. Interestingly, its old rival, Krupp, while in the private enterprise camp, maintains but a tenuous foothold in defence production via a minority holding in MBB, the aerospace producer, its stake until 1986 in Bremen/Bremerhaven shipyards (initially AG Weser, restyled Seebeckwerft) and, more particularly, through a natural affiliate of its former armour interests; namely, the Krupp MaK Maschinenbau tank factory at Kiel. Across the Alps, OTO Melara, now under the auspices of the state EFIM group, earns numerous orders on the basis of its reputation in automatic naval and land artillery, and that is not to overlook its prowess as an MBT producer. Its associate under state control, Breda, also has a fine reputation for anti-aircraft artillery, together with a predilection for aircraft weapons and AFV guns. Only in Switzerland and Sweden do the long-established MIEs of Oerlikon-Bührle (and its subsidiary Contraves) and Bofors (a subsidiary of Nobel Industries) prosper largely independently of government intervention, albeit as users of hybrid traditional-emergent technologies in the air defence combination of missiles and guns. The superpowers remain faithful to either complete state ownership (USSR) or joint public-private control (USA) of ordnance supply. Examples of the latter instance are the procurement of 155mm howitzers from the Rock Island Arsenal on the one hand, whereas, on the other, 20mm Vulcan cannon are bought from GE's Armament and Electrical Systems Department in Burlington, Vermont, and some of the ammunition supply comes from Ford Aerospace.[67]

The exclusion of state arsenals and dockyards from defence production was accomplished most thoroughly in Japan as a direct consequence of post-1945 disarmament. The principal Navy Yards did not escape the axe: Yokosuka was occupied by the US Navy while Kure and Sasebo were turned over to commercial operators for merchant shipbuilding. Reformation of the armed forces in the mid-1950s was entirely incumbent on supply proffered by private firms. Five shipbuilders were licensed by the Maritime Safety Agency in 1953 to recommence the building of warships; namely, MHI, IHI, Mitsui, Sumitomo and Maizuru Heavy Industries (occupying the former state dockyard). Henceforth, government policy ensured that a small core of producers (subsequently joined by KHI and Hitachi, other former wartime MIEs) retained a naval capability through a string of follow-on orders. Yet, throughout the next two decades warship building was generally disparaged by the shipbuilders for the simple reason that they could earn more on their booming merchant ship business than they could from defence fixed-price contracts which scarcely covered production costs. In any event, defence orders for the firms were inconsequential in relation to civil work: for example, MHI captured 33.1 per cent of the Maritime Self Defence Force orders between 1962 and 1966 but that equalled only 0.87 per cent of the group's total sales. Defence dependence was highest with KHI and yet warship sales only amounted to 12.9 per cent of the company's total sales.[68] While hardly qualifying as MIEs, the Japanese firms, nevertheless, imposed a specialised production division on their yards. As the largest, MHI adopted the most comprehensive division of labour: auxiliaries were built at Yokohama, destroyers at Nagasaki, submarines and destroyers at Kobe, and small craft at Shimonoseki. IHI used its Tokyo Number 2 yard for destroyers and its Kure Number 1 yard for patrol craft whereas Mitsui dedicated Tamano to destroyers and Fujinagata to patrol craft. KHI, however, concentrated all its naval work at Hyogo. In a pattern reminiscent of the UK, the Japanese firms have become more defence reliant as merchant shipbuilding demand has evaporated. Rationalisation of merchant shipyards in the 1980s has prompted Sumitomo to set aside its Uraga Works for naval construction, KHI aims to concentrate all its shipbuilding at Sakaide, while Mitsui is concentrating on its Chiba site. The three firms, MHI, KHI and IHI, have also enhanced their defence orientation by strengthening their involvement in military aircraft (the first two) and aeroengine production (IHI).

Table 4.10 identifies the current Japanese warship builders along with their emphasis on destroyer and submarine programmes. It also indicates that the number of warship constructors in Japan compares favourably with the number of builders found in other AICs. In the USA the cutback in warship capacity has been remorseless despite a huge rearmament programme commissioned by the Reagan administration. As recounted elsewhere, even the ostensible merchant yards owe their viability to defence orders. The American Shipbuilding Company, for example, recently completed five T5-class tankers for Ocean Ships of Houston, a company which operates them on long-term charter to the US Navy. Built at Tampa, with components supplied from Lorain and Nashville, their completion left the firm without any merchant newbuilding work at the end of 1986.[69] Incorporation into

Table 4.10 : Principal Shipbuilding MIEs in the AICs

Country	MIE	Main programme(s)
Canada	St John Shipbuilding, St John,NB	City-class FF[1].
West Germany	Bremer-Vulkan	Bremen-class FF
	Thyssen Nordseewerke, Emden	Bremen-class FF/TR-1700 SS
	Blohm & Voss	MEKO-200 FF
	HDW, Kiel	MEKO-200 FF/IKL-1500 SS
Italy	Fincantieri Monfalcone	Sauro-class SS
	Fincantieri Riva Trigoso	Maestrale-class FF/Iraqi FF
	Fincantieri Ancona	Iraqi FF
Netherlands	de Schelde	Standard FF
	Rotterdam Dry Dock	Walrus-class SS
	Wilton-Fijenoord	Sea Dragon-class SS
UK	VSEL, Barrow	Vanguard-class SSBN/Trafalgar-class SSN
	VSEL, Birkenhead	Upholder-class SS/Type 22 FF
	Yarrow, Glasgow	Type 22 FF/Type 23 FF
	Swan Hunter, Newcastle	Type 23 FF/Type 22 FF
USA	GD, Groton (Electric Boat)	Ohio-class SSBN/Los Angeles-class SSN
	Newport News	Nimitz-class CVN/Los Angeles-class SSN
	Ingalls, Pascagoula	Ticonderoga-class CC/Wasp-class LHD
	Bath Iron Works	Ticonderoga-class CC/Arleigh Burke-class DD
	Todd, San Pedro	Perry-class FF
	Lockheed	Whidbey Island-class LSD
	Avondale	Whidbey Island-class LSD
France	Cherbourg DY	SNA-72 SSN
	Brest DY	Richelieu CVN/C70-class DD
	Lorient DY	C70-class DD/FL25-class FF
	CNIM, La Seyne	F2000-class FF
Japan	Mitsubishi, Kobe	Yuushio-class SS
	Mitsubishi, Nagasaki	Hatakazi-class DD
	Kawasaki, Kobe	Yuushio-class SS
	IHI, Tokyo	Hatsuyuki-class DD
	Mitsui, Tamano	Hatsuyuki-class DD
	Sumitomo, Uraga	Hatsuyuki-class DD
	Hitachi, Maizuru	Hatsuyuki-class DD
Australia	Williamstown DY	Perry-class FF

Note: 1. Designations: FF = frigate, SS = submarine, SSN = nuclear submarine,
SSBN = ballistic-missile armed nuclear submarine,
CVN = nuclear-powered aircraft carrier, CC = cruiser,
LHD = helicopter carrier, LSD = assault ship,
DD = destroyer.

larger groups has only temporarily stayed the rationalisation process among US shipbuilders. GD's acquisition of the Bethlehem Quincy yard in 1964 ensured that facility's survival for another two decades, but the end came in 1985 when failure to win a vital order for naval auxiliaries led GD to announce its impending closure with the loss of

6,300 jobs (a pattern putatively echoed by Lockheed which, in mid-1987, offered its shipbuilding division for sale). Bethlehem, incidentally, had retrenched to the extent of concentrating shipbuilding at Sparrows Point while dispensing with most of its ship-repair yards through their sale to the Braswell Corporation in 1982. It was having difficulties enough with its ailing steel operations.[70] Other shipbuilders had fared reasonably well, however. Indeed, the 600-ship naval scheme had inordinately favoured some MIEs: the 1985 new construction budget, for example, assigned fully 73 per cent of the value of the contracts to Newport News, Litton's Ingalls yard and GD's Electric Boat Division.[71] Moreover, as Table 4.10 evinces, the Bath Iron Works, Todd, Lockheed and Avondale yards had not been overlooked. The first of these, Bath, was acquired by the Congoleum Corporation in 1968 (coincidentally, the same year that Tenneco purchased Newport News). Congoleum was a leading manufacturer of cushioned vinyl flooring and had automotive parts activities too. Bath was bought because it showed technical prowess, enjoyed a propriety product line--its destroyer speciality--and dominated a market niche impervious to new competition owing to the high costs of entry.[72] A change of heart on the part of the conglomerate resulted in Bath being offered for sale in 1986. The MIE appeared to have bright possibilities, not least of them being a backlog of orders from the US Navy worth $1.4 billion and, as the 'lead' yard of the Arleigh Burke-class of destroyers, very real prospects of gaining up to 50 follow-on orders.[73] In many respects, Bath's experience typifies the circumstances facing traditional MIEs in the AICs. In short, they have come to play second-fiddle to other, more diverse corporate interests. The defence field has moved on to more rewarding (profitable) technologies and corporate interest has followed suit. In the senescence stage, traditional MIEs are prey to conglomerates that achieve their robustness from a range of activities. Those traditional MIEs not succumbing to conglomerate acquisition may, in the face of certain failure, be saved through nationalisation. While less profitable--if profitable at all--weapons platform production is still an indispensable aspect of defence industry. The postwar Vickers story sums up the options facing traditional MIEs confronting senescence, and a thumbnail sketch of it is hereupon provided.

Confronted with disarmamental instability, Vickers-Armstrong moved to convert part of the

Barrow Works to the manufacture of cement-making machinery, condensers and soap-making equipment. Elswick was steered into making presses while Crayford was turned to the manufacture of petrol pumps. Additionally, a number of civil lines were bought. These included companies making accounting machines, offset lithographic presses, paint grinders and mixers, and filter pumps. Furthermore, the firm boosted its international operations, reacquiring control of Canadian Vickers, purchasing the Cockatoo Docks shipyard in Sydney and other engineering concerns in Australia, and setting up a joint venture cement-making machinery establishment in India. By 1960 it had hived off its aircraft division into the British Aircraft Corporation. By the end of that decade it had lost its steel (armour) interest with the absorption in 1967 of the English Steel Corporation into the state monopoly producer, British Steel Corporation, and its Walker shipyard through it being transferred, in 1968, to Swan Hunter. At that juncture, Barrow was devoted to naval and merchant shipbuilding, nuclear propulsion engineering for submarines, guns, cement plants, pumps, and diesel engines for locomotives. The Dartford explosives factory was closed (though the site was retained for the steel furniture division), as was the Weymouth torpedo factory. Scotswood was phased out with the failure of a venture into tractor making. Elswick continued to depend on MBT production, much of it connected with the creation by Vickers of a tank factory at Madras for the Indian Government. In 1970 the company decided to close the Palmers ship-repair facility (acquired just before World War 2) on the Tyne, and simultaneously sold its computer interests--a progression from accounting machines--to ICL. Paradoxically, therefore, the firm was reliant for up to 70 per cent of its profits on shipbuilding and aircraft on the verge of the nationalisation of these activities in 1977. Conversion into civil lines had been rigorously pursued but with mixed results: office equipment, printing plates, offshore engineering and bottling machinery had proved to be profitable; tractors, chemical engineering, medical engineering and printing machinery had been loss-makers.[74]

Forcibly deprived of its key shipbuilding activity (which, as recounted, became the VSEL subsidiary of BS), Vickers undertook the production of defence items (particularly AFVs), bottling and packaging machinery, printing, hydraulics, machine tools, medical equipment and nuclear engineering. It bought Rolls-Royce Motors in 1980 for £35 million

but rid itself of Canadian Vickers. Notwithstanding
the fact that Vickers Defence Systems was badly hit
by Iranian cancellation of MBT orders, it decided to
designate defence as a core activity and built a new
tank plant at Scotswood to replace the old Elswick
factory. This Newcastle facility was augmented in
1986 by the purchase of ROF Leeds and, thereby, the
assertion of Vickers as the UK's monopoly MBT supplier. Other core activities designated by the firm
offer civil alternatives to defence activity; to
wit, cars (the Rolls marque), lithographic plates,
business equipment and marine engineering.[75] Vickers
PLC is, thus, a very different beast in 1986 from
what it was in 1946, let alone its undertakings of
1914 and the height of the traditional MIE era. It
has survived by balancing declining defence lines
with new-technology defence lines on the one hand,
and diversification--after some trouble--into counter-cyclical civil lines on the other. In so far as
defence activity is concerned, Vickers Defence Systems has shed its mantle of a traditional MIE for
one of a modern MIE.

Imitators and Competitors

In a word, the technology underpinning the traditional MIE has become commonplace. Innovation in
warship weapons platforms and artillery is margi-
nally incremental, but the process technology is
readily transferable and can be mastered by NICs
taking the time and trouble to invest in production
plant. The way was led by the USSR in the massive
industrialisation drives enforced after 1928,
thrusts which rested heavily on defence industry.
Unsurprisingly, then, the USSR is now the world's
major producer of artillery and ordnance and, following from that, is a leading instigator of incremental innovation which can be readily spliced into
the semi mass-production facilities already in
place. More recently, the USSR has also come to
challenge the AICs in warship production. Untouched
by commercial considerations, Soviet naval shipyards
are geared to multi-year planning targets and are
fully able to make use of specialisation at the
plant level. Intimations of the present division of
labour can be got from Table 4.11. From it can be
inferred the fact that the site first developed with
the assistance of the British private MIEs--Nikol-
aiev--has since been expanded into a North complex
concentrating on cruisers and a South complex capable of tackling nuclear-powered aircraft carriers as

well as the steam-powered Kiev-class aircraft carriers. Other yards, sited in the Leningrad vicinity, were also recipients of foreign technology in the Tsarist age, and they continue to play a leading role in Soviet naval shipbuilding. The chief of these, the Baltic Yard, is building the Kirov-class of battlecruiser; the Admiralty Yard is series producing nuclear and conventional patrol submarines, while the Zhdanov Yard specialises in series production of destroyers. It is noteworthy, however, that the state MIEs are scattered, as a deliberate act of policy, throughout the USSR and this is attested by, for example, the naval shipyards at Gorki in the interior, Severodvinsk in the Arctic and Komsomolsk in the Russian Far East.

The USSR perseveres with its own designs but, with the exception of fast attack craft (and larger ships in the special cases of India and China), it has not followed the Western practice of either licensing whole designs to NIC producers or assisting them in formulating their own designs. To date, state-induced international ventures in warship construction have not resulted in NICs jointly participating as co-operative producers in standard designs. The same cannot be said, though, of international state-induced dependent co-producer schemes. Several NICs have established naval shipbuilding capacity in recent years and the entries are continuing apace. One of the most recent newcomers is Thailand. Its Ital-Thai Marine yard in Bangkok, founded with Italian assistance, is constructing a class of tank landing ships to a French Normed design. As a generalisation, new-entry MIEs--such as this Thai example--cut their teeth on licence production before graduating to a modicum of indigenous design capability where they attain the position of dependent co-producers. Thus, India's Mazagon Dock was first prepared--with the assistance of Vickers and Yarrow--to build copies of UK Leander-class frigates and then proceeded to progressively modify them, through several intermediate stages, into the Godavari-class using local design inputs. Through force of circumstances, China has also progressed far along this path. Original copies of Soviet designs were replaced by locally-amended versions after the withdrawal of Soviet technical assistance. We have already highlighted in Chapter 3 the Luda-class, a modification of the Soviet Kotlin-class destroyer, as an instance of limited-indigenous-design capability. To it could be added the Jiangdong and Jianghu-classes of frigate which ultimately derive from the Soviet Riga-class, and the

Table 4.11 : Specialisation of Soviet Naval MIEs

Yard	Responsibility
Severodvinsk	Typhoon-class SSBN Oscar-class cruise missile SSN
Gorki	Sierra-class SSN . Kilo-class SS
Admiralty Yard, Leningrad	Victor III-class SSN Kilo-class SS
Komsomolsk	Victor III-class SSN Kilo-class SS
Nikolaiev South	new CVN class Kiev-class CV
Baltic Yard, Leningrad	Kirov-class battlecruiser
Nikolaiev North	Slava-class CC
Zhdanov Yard, Leningrad	Sovremenny-class DD Udaloy-class DD
Kaliningrad	Udaloy-class DD
Kamysh-Burun, Kerch	Krivak III-class FF
Zelenodolsk	Koni-class FF

Source: Navy International, April 1986 and Jane's Fighting Ships 1986-87.

Ming-class submarine which is an extension of the USSR's Romeo-class. In turn, CSSC has transferred know-how to North Korea's Mayang Do shipyard which not only builds copies of Romeo-class boats but has produced an indigenous improvement of Soviet and Chinese frigates, the Najin-class.

Most NICs, however, are still firmly lodged in the licensee category. Usually, the first-of-class is built by the licensor in the AIC while subsequent vessels are constructed in the NIC under a phased programme of technology transfer. Several Latin American MIEs conform to this pattern. Brazil's Arsenal de Marinha in Rio de Janeiro is a case in point. It acquired sophisticated frigate technology by producing the last two of the Niteroi-class after the first four vessels had been completed in the UK. Similarly, it builds West German Type 209 submarines and Italian corvettes under technology transfer schemes. In fact, Brazil has gained sufficient confidence in naval construction to order warships from a domestic private firm. Formerly a subsidiary of a Dutch merchant shipbuilding concern, Verolme do Brasil is building two corvettes to supplement comparable vessels emanating from the Arsenal shipyard. Argentina's AFNE and Astilleros Domecq Garcia function according to similar criteria, while Peru's SIMA yard at Callao is completing the country's buy of Lupo-class frigates after the first two were supplied by the CNR Riva Trigosa yard in Italy. Blohm + Voss, a member of the Thyssen Nordseewerke shipbuilding group, is something of a specialist in arranging such technology transfer ventures and serves as a typical example of a 'lead prime', licensing its designs to new-entry MIEs which function as dependent co-producers. Currently, its MEKO-200 frigate design is under construction at the Turkish Naval Shipyard at Gölcük and the West German firm is itself building two (with another from Howaldtswerke-Deutsche Werft) for Portugal. Six of its MEKO-140 frigates are the subject of a collaborative programme with AFNE of Argentina.

On the receiving end of technology transfer are such longstanding MIEs as Australia's Williamstown Dockyard and Spain's Bazán, both state enterprises (although, reportedly, the former is destined for privatisation).[76] Assisted by Todd Shipyards, the first is building two FFG7-class frigates despite completing its last ship--a UK design--in 1971. Employing 14,000 people, the second is a more ambitious undertaking. Part of the Instituto Nacional de Industria (along with CASA, an aircraft MIE), Bazán is rooted to a tradition of licence production (for example FFG7 frigates) combined with some qualitative improvements which push it into the category of a dependent co-producer (e.g. aircraft carrier work). Indeed, Bazán has gathered technology from a wide variety of donors since its emergence in 1947 as a replacement of the old SECN organisation originally inspired by a group of British MIEs. It builds submarines based on the French Agosta-class at Cartagena, frigates of the American type at Ferrol and F-30 frigates at both yards. The last is a ship of convoluted pedigree, being based on the Portuguese Joao Coutinho-class which was designed by Blohm + Voss and engined by diesels designed by MTU. The fast attack craft emanating from its La Carraca yard owe their genesis to the West German firm of

Lürssen.   Bázan has put this bought-in expertise to
good use,   selling fast attack craft   to Argentina,
Mexico, Congo, Morocco, Egypt and Mauritania; Agosta-class submarines to Egypt;   and F-30 frigates to
Egypt and Morocco. Despite their foreign parentage,
the submarines are produced to the extent of 67 per
cent from Spanish-supplied inputs while the frigates
are so sourced to the degree of 83 per cent.[77] Other
challengers to the AICs bobbing up in export markets
are Singapore and South Korea.   The former is supplying, courtesy of the Singapore Shipbuilding &
Engineering concern, the Indian Coast Guard with six
200-tonne patrol vessels under a $30 million con-
tract.   The latter, through Korea Tacoma of Masan,
has supplied PSMM-5 fast attack craft to Indonesia
as well as tank landing ships--and, in so doing,
subscribes to the category of co-operative prime in
partnership with the US Tacoma enterprise (although,
ironically, the American firm declared bankruptcy at
the end of 1985). Diversifying from its position as
one of the world's largest merchant shipbuilders,
Hyundai is a new-entry naval supplier, delivering
improved versions of US-designed tank landing ships
to Venezuela and Argentina.   Another NIC with good
export credentials is Israel.   The Haifa yard of
state-owned Israel Shipyards was instrumental in
setting up Sandock Austral of South Africa as a
dependent co-producer of fast attack craft under an
international state-induced agreement.   In spite of
such export successes, Israel Shipyards stands testimony to the fact that NIC warship enterprises are
not immune to market failure. Publicly revealed to
be unprofitable as a result of declining defence
orders, the Haifa yard is kept operational solely on
the basis of government hand-outs and has suffered
the ignominy of being placed under a temporary
receiver. The yard's 720 jobs are put in jeopardy
and the Israeli Government is considering selling
the facility to private (perhaps US) interests.[78]
   Weapons platforms of the size and complexity of
warships still impose some entry constraints--and
questions of survival once created--on aspiring
defence producers, as the Israeli case attests. Yet,
fewer barriers to entry confront NIC enterprises
wishing to set up ordnance plants. Modern small arms
manufacture is now virtually universally undertaken
by NICs; indeed, its antecedents in countries such
as India reach far back into colonial times. Artillery is produced by a more restricted number of countries, but those NICs that have plunged into its
manufacture have arrived at the point where they can
truly compete on equal   terms with traditional

195

Table 4.12 : Contemporary Artillery Manufacturers

Indigenous design and production	Derivative or licence producers
(A) Self-Propelled Guns and Howitzers	
Czechoslovak state arsenals	Chinese state arsenals
GIAT, France	Abu Zaabel, Egypt
Rheinmetall, West Germany	Komatsu, Japan
Soltam, Israel	Taiwan state arsenals
OTO Melara, Italy	
Nihon Seiko Jyo, Japan	
Armscor, South Africa	
AB Bofors, Sweden	
Soviet state arsenals	
VSEL, UK	
Bowen-McLaughlin-York, USA	
(B) Towed Guns	
Czechoslovak state arsenals	CITEFA, Argentina
Oy Tampella, Finland	Australian Ordnance Factories
GIAT, France	Voest-Alpine, Austria
Indian Ordnance Factories	Chinese state arsenals
Rheinmetall, West Germany	Abu Zaabal, Egypt
OTO Melara, Italy	Nihon Seiko Jyo, Japan
Soltam, Israel	Kia Machine Tool, South Korea
Israel Military Industries	RDM, The Netherlands
Singapore Technology Corp.	Taiwan state arsenals
Lyttelton Engineering, South Africa	Hsing Hua, Taiwan
Fabrica de Artilleria de Sevilla, Spain	Yugoslav state arsenals
AB Bofors, Sweden	
Soviet state arsenals	
RO Nottingham, UK	
Rock Island Arsenal, USA	

Source: Derived from information in Jane's Armour and Artillery 1985-86.

producers in the AICs. Table 4.12 draws up a list of the world's principal suppliers of artillery pieces outside of the anti-aircraft and naval arenas. The more sophisticated self-propelled variety are designed by a handful of MIEs located, for the most part, in the AICs and the Soviet bloc (not least from the successor to Skoda in Czechoslovakia). Notable among them, though, are two NIC producers; namely, Soltam of Israel and Armscor of South Africa. Three other NICs (and Japan) produce self-propelled guns to foreign designs: US in the case of Taiwan and Soviet in the cases of China and Egypt. Less sophisticated guns of the towed variety spring from a larger set of producers. Indigenous design of these weapons is carried out by such disparate producers as Tampella in Finland, the Indian Ordnance Factories, the Singapore Technology Corporation and Spain's Seville arsenal. In the main, technology transfer is from the AICs to the NICs. Thus, Argentina's CITEFA 155mm howitzers derive from French F3 designs, the Australian Ordnance Factories 105mm gun was designed by RO in the UK, Austria's weapon is a Belgian design, while the Japanese are building copies of the FH-70 155mm howitzer jointly designed by RO, Rheinmetall and OTO Melara. Israeli inspiration is behind the efforts of Taiwan while that of the USA is responsible for South Korean, Dutch and Yugoslav output. The USSR continues to influence China and Egypt via the 'reverse engineering' process, that is to say, through unlicensed manufacture of Soviet designs and incremental improvements of them. The presence of Israel in the ranks of those promoting technology transfer is an augury of NIC penetration of such markets, just as the role of the USSR in sourcing ordnance production is expressive of an earlier successful incursion into markets once monopolised by the traditional MIEs of the AICs.

SUMMARY

The culmination of several significant advances in civilian technology coalesced to impact on defence technology in the middle years of the 19th century. Innovations in metallurgy not only influenced weapons, providing especially the enabling technology which secured the development of the breech-loading rifled artillery piece, but also dramatically shaped the platforms which mounted the weapons. The ironclad ushered in the age of metallurgical expertise which reached its pinnacle with the 'dreadnought' battleship of the early 20th century. The 'dreadnought' was a large and complicated weapons platform requiring massive investment in big guns, armour protection and marine engineering capability. Its rise was paralleled by the evolution of what we have termed the 'traditional' MIE. Initially skilled in one aspect of the new technology, gun-making, armour-making or shipbuilding, the MIE increasingly came to straddle all of them. Indeed, MIEs unable or unwilling to pursue such integration strategies were, in general terms, steadily reduced to secondary status. Many of the increasingly marginalised MIEs were longstanding state undertakings--arsenals or dockyards--which, for a variety of reasons, failed to keep pace with the technical innovation and organisational aggrandizement that came to typify the leading private MIEs. Some state enterprises contented themselves with the reinforcement of their specialisations in certain niches: Portsmouth Dockyard with 'dreadnoughts' and the Enfield arsenal with small arms are cases in point. Others prevailed as a result of state nurturing and protection; for example, the Arsenal de Brest. It was the likes of Vickers, Armstrong, Krupp and Bethlehem, however, which supplied the weapons systems across the board: ships, guns, armour and a welter of ancillary equipment.

World War 1 served as a watershed for this first generation of traditional MIEs. Peak production of ordnance and warships was followed by hesitancy in defence budgets in general and questions as to the suitability of traditional weapons systems in particular. Rivalled by air power and mechanised land warfare claimants, the traditional MIEs were obliged to diversify assets that had, in any event, been threatened by wholesale postwar cutbacks in defence demand. Rationalisation of ordnance, armour, shipyard and gun plant was accompanied by modest diversification into such defence fields as aircraft and AFVs, not to speak of experimental inroads into many and varied civil lines. Conversion attempts were redoubled after the production respite offered by World War 2. Moreover, heightening the problems of declining peacetime domestic demand was the steady encroachment on export markets by newcomers directly stimulated by that war and its post-colonial aftermath. New entries from MIEs in China, India, Israel and elsewhere posed threats to the market outlets of AIC producers submerged in the senescence stage of the product life-cycle. Ready acquisition of mature technology supported the reinvigoration of Soviet defence production which tended to undercut those markets even further. Consigned to a future of apparent secular decline, the traditional MIEs in the AICs either succumbed to permanent receivership as de facto (if not always, de jure) state enterprises or became divisions of multidivisional private conglomerates. Recently, a bout of competitiveness in AIC defence procurement has required some organisational changes, most visible in the UK's drive to privatise those traditional MIEs--shipbuilding, ship-repair and ordnance--that had fallen under the state tutelage. In the meantime, aspiring NICs continue to press for technology transfer schemes which allow for their entrance into weapons platform production; a situation which is certain to intensify the international competition in the field of traditional MIE activities.

NOTES AND REFERENCES

1. The outcome, of course, of the new international division of labour in defence production introduced in Chapter 1.
2. Cited in J. D. Scott, Vickers: a history, (Weidenfeld & Nicolson, London, 1962), p.47.
3. C. Trebilcock, The Vickers brothers: armaments and enterprise 1854-1914, (Europa Publications, London, 1977), pp.7-8.
4. Trebilcock, Vickers brothers, p.8.
5. J. R. Hume and M. S. Moss, Beardmore--the history of a Scottish industrial giant, (Heinemann, London, 1979), p.45 and pp.72-7.
6. M. Kaldor, The baroque arsenal, (Hill & Wang, New York, 1981), p.39.
7. R. Humble, The rise and fall of the British navy, (Queen Ann Press, London, 1986), p.32.
8. B. Pool, Navy Board contracts 1660-1832, (Longman, London, 1966), p.86.
9. The figures in Table 4.1 are calculated from the listings provided in J. J. Colledge, Ships of the Royal Navy, (2 vols., David & Charles, Newton Abbot, 1969 and 1970).
10. The former is often attributed to the UK Admiralty's insistence on mechanised dockyard apparatus while the latter stems from the efforts of US entrepreneurs in the gun-making trade who were responding to specifications issued by the US Army. See M. Pearton, The knowledgeable state, (Burnett Books, London, 1982), p.40.
11. Pearton, Knowledgeable state, p.44.
12. A point of view expressed in the 1854 Report from the select committee on small arms, an official body of enquiry set up to examine the shortcomings in small arms supply exposed by war preparations. See E. Ames and N. Rosenberg, 'The Enfield Arsenal in theory and history' in S. B. Saul (ed.), Technological change: the United States and Britain in the nineteenth century, (Methuen, London, 1970), pp.99-119. It has been estimated that the government assigned 75 per cent of all rifle orders between 1864-78 to Enfield, the balance going in the main to BSA (founded 1861) and the London Small Arms Company: two firms which were equally adept at absorbing US principles of interchangeable parts production. Note S. B. Saul, 'The market and the

development of the mechanical engineering industries in Britain, 1860-1914', pp.141-70 in the same volume.
13. H. Lyon, 'The relations between the Admiralty and private industry in the development of warships' in B. Ranft (ed.), Technical change and British naval policy 1860-1939, (Hodder & Stoughton, London, 1977), pp.37-64.
14. German copies of the Maxim gun were made at the Spandau factory while a plant was set up at Tula in 1905 to make them for the Russian military.
15. W. Hornby, Factories and plant, (HMSO, London, 1958), p.6.
16. P. Barry, Dockyard economy and naval power, (Sampson Low, London, 1863), pp.209-57.
17. B. F. Cooling, Gray steel and blue water navy, (Archon Books, Hamden, Conn., 1979) and W. R. Herrick, The American naval revolution, (Louisiana State University Press, Baton Rouge, 1966). Note, the introduction of nickel into armour triggered the formation of two cartels: the French-led Le Nickel which obtained the metal from New Caledonia in the Pacific and fed it to the European armour makers, and Inco which extracted the ore from Ontario and supplied the American armour firms. After 1902, both Inco and US Steel (the erstwhile Carnegie) were controlled by that arch-entrepreneur J. P. Morgan. See J. R. Lischka, 'Armor plate: nickel and steel, monopoly and profit' in B. F. Cooling (ed.), War, business, and American society, (Kennikat, Port Washington, NY, 1977), pp.43-58.
18. J. Y. Lancaster and D. R. Wattleworth, The iron and steel industry of West Cumberland: a historical survey, (British Steel Corporation, Workington, 1977), pp.82-90.
19. A. Grant, Steel and ships: the history of John Brown's, (Michael Joseph, London, 1950), p.50. Eventually, the two firms were to amalgamate in 1930.
20. D. K. Brown, A century of naval construction: the history of the Royal Corps of Naval Constructors, (Conway Maritime Press, London, 1983), pp.35-6.
21. K. J. W. Alexander and C. L. Jenkins, Fairfields: a study of industrial change, (Allen Lane, London, 1970). The company survives as BS subsidiary, Govan Shipbuilders Ltd.
22. Brown, Century of naval construction, p.90.
23. T. A. Brassey, Naval annual 1892. These figures contrast with the average of £1.25 million prevailing from 1863 to 1881.
24. E. Wilkinson, The town that was murdered:

the life-story of Jarrow, (Victor Gollancz, London, 1939), pp.110-1. See also issues of Fairplay for 18 September 1891, 13 November 1891, 27 May 1892, 30 September 1892, 20 September 1900 and 19 September 1901.
25. J. F. Clarke, Power on land and sea: 160 years of industrial enterprise on Tyneside, (Smith Print Group, Newcastle, 1977), pp.50-65.
26. Brown, Century of naval construction, p.92.
27. See Shipbuilder, May 1922, p.322.
28. Cooling, Gray steel and blue water navy, p.41 ff.
29. See Fairplay, 20 November 1986, p.57.
30. R. W. Johnson, The making of the Tyne, (Walter Scott, London, 1895) and D. Dougan, The history of North East shipbuilding, (George Allen & Unwin, London, 1968).
31. M. H. Dodds, A history of Northumberland, vol. XIII (Andrew Reid, Newcastle, 1930), pp.253-5 and R. J. Irving, 'New industries for old? some investment decisions of Sir W. G. Armstrong, Whitworth & Co Ltd, 1900-1914', Business History, vol. 17 (1975), pp.151-68. Armstrongs were consistently earning profits of £250,000 and more in the early-1890s, rising to £650,000 by the end of the decade. Its capital was increased to £3 million in 1895 and £4.2 million in 1901.
32. S. Pollard, 'The economic history of British shipbuilding 1870-1914', unpublished PhD thesis, University of London, 1951, p.452.
33. Scott, Vickers, pp.57-85 and Trebilcock, Vickers brothers, pp.154-5.
34. Hume and Moss, Beardmore, p.40 ff.
35. M. R. Smith (ed.), Military enterprise and technological change: perspectives on the American experience, (MIT Press, Cambridge, Mass., 1985), p.8.
36. Cooling, Gray steel and blue water navy, p.199.
37. B. Dethomas, 'Creusot-Loire--end of a 200-year industrial saga', Manchester Guardian Weekly, (15 July 1984), p.12.
38. R. D. Burns, The international trade in armaments prior to World War II, (Garland, New York, 1972), p.58.
39. R. J. Q. Adams, Arms and the wizard, (Texas A & M University Press, College Station, 1978), p.100.
40. Hornby, Factories and plant, p.9.
41. P. Connon, An aeronautical history of Cumbria, Dumfries and Galloway region, Part 2: 1915 to 1930, (St Patrick's Press, Penrith, 1984),

pp.109-14.
42. L. Hannah, The rise of the corporate economy, (Methuen, London, 1976), pp.118-9. Note, by
1930 only Vickers survived--at rank twelve--in the
top 50.
43. Wilkinson, Town that was murdered,
pp.116-7.
44. D. A. Hounshell, 'Ford Eagle Boats and mass
production during World War I' in Smith, Military
enterprise, pp.175-202.
45. C. Trebilcock, 'British armaments and European industrialization, 1890-1914', Economic History
Review, vol. 26 (1973), pp.254-72 and the same
author's The industrialization of the Continental
powers 1780-1914, (Longman, London, 1981).
46. U. Kobayashi, Military industries of Japan,
(Oxford University Press, New York, 1922).
47. C. Trebilcock, ' "Spin-off" in British economic history: armaments and industry, 1760-1914',
Economic History Review, vol. 22 (1969), pp.474-90.
48. Sir W. G. Armstrong, Whitworth & Co (Shipbuilders) Ltd retained the Walker, Dobson and Tyne
Iron shipyards while Sir W. G. Armstrong, Whitworth
& Co (Engineers) took charge of propelling machinery. By the same token, all locomotive work at
Elswick and Scotswood and machine tool work at Openshaw was excluded from the merger.
49. D. Todd, The world shipbuilding industry,
(Croom Helm, London, 1985), p.238 and p.334.
50. Scott, Vickers, p.187. Note, John Brown
had withdrawn from gun forging work by this date.
51. Burns, International trade in armaments,
p.95 and pp.259-60.
52. A. W. Saville, 'The naval military-industrial complex, 1918-41' in Cooling (ed.), War, business, and American society, pp.105-17.
53. Bofors 40mm cannon, for instance, were
built under licence in the UK, USA and Canada. In
the USA, no fewer than 34,000 were built during the
war by Chrysler, Firestone and the Pontiac Division
of GM.
54. E. R. Zilbert, Albert Speer and the Nazi
Ministry of Arms, (Associated University Presses,
East Brunswick, NJ, 1981), pp.134-6.
55. Hornby, Factories and plant, pp.132-3.
56. Specialisation was not carried to extreme
lengths. Thornycroft, for example, also made torpedo
parts, Bren gun carriers and artillery pieces (at
Basingstoke), not to overlook its leading role as a
supplier of fast attack craft (from its Hampton-on-
Thames works).
57. Brown, Century of naval construction,

p.295.
  58. J. Soppelsa, Géographie des armements, (Masson, Paris, 1980), pp.191-2.
  59. Fifth report from the committee of public accounts, House of Commons paper 556, session 1975-76.
  60. Thirty-fifth report from the committee of public accounts, House of Commons, session 1984-5: design and procurement of warships, Ministry of Defence, p.x.
  61. As reported in The Engineer, 1 November 1984, p.28.
  62. In 1986 Harland & Wolff won an oil replenishment vessel order potentially worth 2,000 jobs for itself and for Yarrow and Racal Electronics over four years. The company's market for large tankers and bulk carriers in the merchant trades had evaporated.
  63. BS has already released some ship-repair yards into a private enterprise system in which they are inordinately reliant on MoD work. Falmouth Shiprepair Ltd set the tone for competitive tendering within the Royal Dockyards by winning a £3.5 million contract to convert a merchant products carrier into a naval auxiliary. Humber Shiprepairers of Immingham paved the way with respect to competitive tendering for submarine refits. See Fairplay, 19 June 1986, p.32. Ironically, privatisation of the management of the Royal Dockyards was accomplished only at the expense of firms such as Humber Shiprepairers; for, the new Dockyards managers were accorded three years of exclusive naval contracts as an incentive for undertaking the task. Refer to The Engineer, 19 February 1987, pp.30-1.
  64. Parties under consideration for Devonport Dockyard included consortia linking Brown & Root and Vickers PLC, Foster Wheeler and A & P Appledore (a shipbuilding consultancy already operating the former Gibraltar Dockyard), and a management team. Three parties also bid for Rosyth Dockyard: Babcock International combined with Thorn EMI Electronics, Balfour Beatty and the Weir Group, and Press Offshore. See D. Wettern, 'The dockyard battle', Navy International, vol. 91 (1986), pp.763-5. In point of fact, the Brown & Root consortium prevailed for Devonport and the Babcock-led group for Rosyth.
  65. BAe gained RO for a price of £190 million. As part of the purchase agreement, it had to refrain from entry into MBT work and hence, competition with Vickers, for five years. Refer to Jane's Defence Weekly, 30 May 1987, p.1080.
  66. Small-arms work is to be transferred to

Nottingham. Note, Waltham Abbey has also been closed. Refer to Jane's Defence Weekly, 22 August 1987, p.343.

67. Many US Government arsenals are operated as GOCO plants. For example, Chamberlain National runs the Scranton and Mississippi Army Ammunition Plants as well as its own Waterloo, Iowa, and New Bedford, Massachusetts, factories.

68. J. E. Auer, The postwar rearmament of Japanese maritime forces, 1945-71, (Praeger, New York, 1973), p.235.

69. As mentioned in Fairplay, 11 December 1986, p.24.

70. See Business Week, 4 August 1986, p.25. Bethlehem had to write off 5.4 million tons of steel capacity at a cost of $1.8 billion over the years 1977-82.

71. As reported in USA Today, 4 October 1985, Section B, pp.1-2.

72. R. L. Kuhn, Mid-sized firms: success strategies and methodology, (Praeger, New York, 1982), p.201.

73. See Business Week, 23 June 1986, p.47. Companies expressing interest in Bath Iron Works included GE, Hewlett-Packard and MD.

74. Scott, Vickers and H. Evans, Vickers: against the odds 1956-1977, (Hodder & Stoughton, London, 1978).

75. As recounted in various issues of the journal The Engineer. For example, see copies of 3 July 1980, 2 October 1980, 10 March 1983, 9 February 1984, 1 March 1984 and 24/31 July 1986.

76. See Jane's Defence Weekly, 23 May 1987, p.996 and 19 September 1987, p.639. By late-1987, five groups had put in bids to take over the yard.

77. E. Loose-Weintraub, 'Spain's new defence policy: arms production and exports' in SIPRI, World armaments and disarmament, (Taylor & Francis, London, 1984), pp.137-49.

78. As revealed in Navy International, vol. 91 (1986), p.249. It is worth noting that a modern MIE, Bet Shemesh Engines Ltd, is also under threat of collapse despite being a jet aeroengine maker and a majority-owned state undertaking. Evidently, failure of NIC state MIEs is a phenomenon being pioneered by Israel. Bet Shemesh, incidentally, will see its government stake transferred to industrialist Stef Wertheimer (Pratt & Whitney, however, retaining its 42 per cent holding) while, simultaneously, $93 million in debts is written off by the state. See Aviation Week & Space Technology, 31 August 1987, p.27.

Chapter Five

THE MODERN MIE

In large measure, the modern MIE is an offspring of
World War 1. As was evidenced through Table 1.5,
that conflict was responsible for ushering into
existence a host of 'infant' aircraft producers.
Moreover, Winston Churchill's famous act of invoking
motor vehicles and agricultural machinery producers
to come up with the tank was also, of course, part
of the technological revolution in weapons systems
prompted by the Great War. However, its immediate
aftermath of demobilisational instability and the
associated precipitous decline in demand for aircraft and AFVs almost had the effect of excising
these flowering entrepreneurial initiatives. It is
no exaggeration to aver that the modern MIE was a
precarious entity throughout the 1920s. Nevertheless, the rise of Fascism, not to speak of Stalin's
military industrialisation, and the belated rearmament of the Democracies in the 1930s conspired not
only to revive the aircraft and tank manufacturers,
but to accord them a status equalling that of the
traditional MIEs. The onset of another global war
with its onus on mobility and mechanised operations
was enough to give precedence to the modern MIEs in
the defence industry scheme of things. Its culmination with the introduction and use of nuclear weapons established the bomber at the pinnacle of strategic weaponry and, more tellingly for our purposes,
accorded corresponding salience to those modern MIEs
capable of furnishing bombers. The subsequent era,
characterised by intermittent Cold War between the
superpowers, has emphasised the prominence of air
defence--even when strategic bombers were supplanted
by ICBMs and, later, SLBMs--and tactical air forces
have been the cornerstone of such defence. Since
World War 2, then, aircraft weapons systems have
reigned supreme in defence thinking (and are only
now being challenged by missiles and EW).

Consequently, the modern MIE has both prospered signally and dominated national defence industries.
     The object of this chapter is not to embellish
the record of alternating growth and decline of the
aircraft industry in response to the dictates of the
wave-cycle.   That purpose has been accomplished
elsewhere (recall Chapter 1). By the same token, it
does not set out to amplify the details surrounding
the evolution of aircraft and MBT technology in conjunction with enterprise advancement: that, too, has
been dealt with elsewhere.[1] Rather, it aims to consider the modern MIE within the terms of reference
expounded in Chapter 3. In other words, the object
is to unveil facets of defence-industry organisation
peculiar to the modern MIE.   Before embarking on
that mission, however, a further qualification must
be proffered; namely, it is necessary to own to a
pre-occupation herein with only one branch of the
modern MIE. Not to put too fine a point on it, as
aircraft manufacture and AFV manufacture propagate
together the characteristics of the modern MIE, and
as the former is by far the more salient in defence
industry terms, the chapter will lay greater store
on the aircraft enterprises rather than those geared
to MBT production. While this procedure does not
quite give full measure to the range of modern MIE
operations, it captures the gist of them and, in
view of space limitations, obviates the need to venture into the realm of the automotive industry: a
topic that would tax the confines of a single book.
This is not to say that AFV manufacturers are
ignored: to dispel accusations of omission, full
recognition is accorded to those firms which have
encroached on AFV manufacture from other branches of
defence production.
     Qualifications aside, the evolutionary approach
is not eschewed altogether. An evolutionary under-
tone persists throughout the chapter by virtue of
the initial concern for the appearance of modern
MIEs in the AICs, the subsequent contemplation of
their development in various manifestations in those
same countries, and the concluding interest in their
diffusion into the NICs. Evidently, the issue of
modern MIE formation serves as the point of departure for the chapter, and the aircraft industry's
life-cycle is illustrated by way of historical examples drawn, in the main, from European and American
experience. Essentially, the maturity of the industry is demonstrated via the technological progression from licence producer, through the categories
of incremental improvement and limited indigenous
design, to arrive at enterprises deserving of the

fully-indigenous design appellation. In the first instance, the progression is addressed from a historical viewpoint and, in this vein, the chapter deliberates the circumstances surrounding the motivations that have arisen to nudge traditional MIEs into participation in aircraft and AFV manufacture. This acts to delimit the structural factors mandating industry development and leads, as a matter of course, to contemplation of the situation applying at the present day. The upshot, in fact, is a review of the contemporary modern MIE located in the AICs in which examples of individual enterprises are invested with the attributes of lead and dependent producers. At that juncture, we come at length to the activities of modern MIEs in the NICs. Their performance is set off against the record of the contemporary modern MIEs in the AICs. The continuing technological dependency of many modern MIEs on patrons and donors is finally brought home in the section dealing with offsets, a theme providing a fitting finale for the chapter.

FORMATION OF ENTERPRISES

As hinted in the foregoing account, the 'conception' stage of the aircraft industry life-cycle just predated the occurrence of World War 1. The innovators and tinkerers, however, were soon caught up in the upsurge in military demand and the stages of 'birth', 'childhood' and 'adolescence' were telescoped into four war years in response to galvanised production requirements. A. V. Roe & Company, for example, had begun when the founder started to build his own aeroplane designs in a Manchester basement shop in 1910. War brought expansion through usage of a large plant at Newton Heath. All told, the firm had built no fewer than 3,696 Avro 504s by the Armistice, along with two model 529s and 25 model 521s.[2] Frederick Handley Page, in turn, began building his own aircraft in 1910 and two years later had his workshops located in converted stables in Cricklewood, London. The war was instrumental in converting Handley Page Limited into a heavy bomber producer and a large plant at Cricklewood was obtained for the purpose of constructing these machines. The firm turned out 40 of its preliminary O/100 bombers, 279 of the improved O/400s and 21 of the giants of the age, the model V/1500. Start-up aircraft enterprises undertaking construction of their own designs also embraced the Aircraft Manufacturing Company of Hendon (founded 1912), Blackburn Aeroplane of Leeds (formed 1914), British & Colonial (later Bristol) Aeroplane (founded 1910), Short Brothers of Rochester (founded 1908) and the Sopwith Aviation Company of Kingston-on-Thames (founded 1912 and reconstituted as Hawker Engineering in 1920).

Far more common, however, were new entries building aircraft under licence to these firms or to the state's Farnborough-based Royal Aircraft Factory. The new entries fell into three groups: those which were genuine start-ups, those switching from civilian product lines to warplane manufacture and, finally, those representing diversification of traditional MIEs into aircraft (and AFV) manufacture. Abstracting from the British experience yet again, one can list the Alliance Aeroplane Company of Hammersmith, the Gosport Aircraft Company, Hewlett & Blondeau of Leagrave, the London Aircraft Company and the Air Navigation Company of Addlestone as examples of aircraft industry start-ups. Of this set, Alliance contracted to build ten Handley Page V/1500s, 200 Aircraft Manufacturing Company (Airco) DH.10s and 350 Airco DH.9s. Diversions from the automotive industry included Austin, Crossley, Daimler, Darracq (subsidiary of a French car maker), Humber, Mann Egerton, Napier & Sons, Siddeley-Deasy, Standard Motor, Sunbeam and Wolseley. The Birmingham factory of Austin concentrated on Royal Aircraft Factory designs, building no fewer than 52 RE.7 and 300 RE.8 reconnaissance machines as well as 1,550 of the SE.5a fighters. Conversely, the Wolverhampton factory of Sunbeam focused on Shorts and Avro designs, making 85 of the former and 610 of the latter. A host of other manufacturers entered the fray. Thus, the Norwich firm of Boulton & Paul diversified from building materials, Parnall of Bristol switched from woodworking, Brush of Loughborough diverted from electrical engineering, and the Birmingham Carriage, Metropolitan Waggon and Vulcan companies transferred from railway engineering. The Metropolitan Waggon Company of Birmingham, for instance, was responsible for 100 Handley Page O/400 bombers. Equally noteworthy was the fact that the traditional MIEs were not tardy in entering into both aircraft and tank production. While the significant involvement of Vickers in AFV design and manufacture was a symptom of postwar diversification as the state at first reserved such work for its own arsenals, the firm's aircraft division became a force to be reckoned with during the war. Not only did it manufacture large numbers of Farnborough-designed aeroplanes, but it formulated and built its own scout and bomber aircraft. In the latter respect, it was

The Modern MIE

emulated by Armstrong Whitworth and Beardmore (both sources of battleships) and J. Samuel White (the Cowes destroyer builder). In contrast, other naval suppliers such as Barclay Curle, the Coventry Ordnance Works, Denny, Fairfield, Harland & Wolff and Stephen preferred to produce aircraft under licence.

DIVERSIFICATION INTO THE NEW SECTOR

The British were not alone in co-opting ordnance and shipbuilding enterprises into aircraft supply at this time. The principal gun-maker for the Austro-Hungarian empire, the Skoda Works, was instrumental in setting up the Oeffag aircraft works in Vienna Neustadt and, indirectly, the Ufag aircraft works in Budapest.[3] Similarly, the Otto Werke of Munich was the aircraft manufacturing subsidiary of MAN, a large producer of diesel engines for submarines. Italian warshipbuilder Ansaldo was unexcelled in this respect: acquiring the aircraft enterprises of Pomilio, SIT and SVA; all collocated in Turin. Automobile manufacturer Fiat undertook both AFV and aircraft manufacture. In the latter sphere it was represented by SIA at Turin for airframe production and Fiat San-Georgio at Muggiano for aeroengine supply. In truth, many traditional MIEs withdrew from aircraft manufacture on the termination of hostilities. Several, however, stayed the course. After 1924, for example, Vickers vied with the ROF in UK tank design and production. MAN, likewise, became a major player in German MBT production. With the onset of rearmamental instability in the 1930s, they were actively encouraged by the state to re-enter the field of MBT and aircraft manufacture. In so far as the first is concerned, Vickers was joined by Harland & Wolff, Rolls-Royce, English Electric (the successor to the Coventry Ordnance Works) and a number of automotive firms in British tank production. In addition, the motor vehicles makers that had cut their teeth on aeroplane construction in the Great War were successfully induced to tool up for aircraft manufacture once again. Under the shadow factories scheme, they were to go on to produce a high proportion of the succeeding wartime aircraft output (recollect Table 1.6). The Americans followed suit, co-ercing their much larger automotive industry into licence production of aircraft designed by the lead primes. Ford, for example, built a huge plant at Willow Run, Michigan, expressly to turn out large numbers of Consolidated (later GD) B-24 Liberator bombers. The UK branch of the company, incidentally, was a major player in the Rolls-Royce Merlin aeroengine production scheme. GM, for its part, concentrated on US Navy needs, and its Eastern Aircraft organisation made such types as the Grumman Hellcat fighter and Grumman Avenger torpedo-bomber. Germany, meanwhile, opted to entice the engineering concerns into its aircraft production effort. From the shipbuilding industry came Blohm + Voss with a Hamburg factory and AG Weser which contributed plants at Einswarden and Lemwerder.[4] Locomotive maker Henschel opened an aircraft factory at Schönefeld, railway-carriage maker Gothaer donated two plants at Furth and Gotha, while transport-equipment maker Ago added its Oschersleben plant.

As mentioned in Chapter 3, motor vehicles producers have consistently cemented fruitful ties with defence industry. From the ranks of British car and truck makers a veritable AFV industrial complex was forged in World War 2. AEC, a manufacturer of truck and bus chassis, turned its hand to Matador gun tractors. Vauxhall Motors, a manufacturer of cars for the mass market, redesigned the Churchill tank as the improved Black Prince MBT. Other incremental improvements on existing designs included Leyland Motors' revamping of the Cromwell into the Centaur tank, and the progressive development of the cruiser tank (e.g. the Crusader) by Nuffield Mechanization. The US record was comparable. The Cadillac Motor Division of GM worked with the government's Ordnance Department in designing the M24 Chaffee light tank, FMC made amphibious tractors for the Marine Corps, whereas Ford embarked on mass production of Jeeps and the GAA V-8 standard tank engine, not to speak of its series production of M4 Sherman MBTs and M8 armoured cars (the last a product of its own design offices). A typical entry, Caterpillar Tractor, established a subsidiary--the Caterpillar Military Engine Company--at Decatur, Illinois, and proceeded to manufacture diesel engines for the M4 tank.[5] Perhaps the supreme example of an automotive firm transforming itself into a modern MIE is Rolls-Royce. The company had emerged from the Great War as a significant aeroengine producer, the bellwether of aeroengine design, in fact, and retained military aviation interests--albeit on a much smaller scale--throughout the phases of demobilisational instability and peacetime equilibrium which followed. Anticipating events, as it were, the firm opened a flight-test centre at Hucknall in 1934. By the start of World War 2 its employment had doubled to 12,500 as a consequence of the establishment of shadow factories at Crewe and Glasgow (Hillington). These units complemented the main plant at Derby in aero-engine manufacture. At its peak, Rolls-Royce employed 55,600 and, not satisfied with producing aeroengines, the firm had started a Nottingham factory to produce tank engines as well.[6]

A renewed bout of disarmamental instability after 1945 induced several automotive firms to withdraw from defence production. Others remained, despite sharp shortfalls in activity (e.g. Alvis, Allison and Renault). To a considerable degree, though, the dampening effects of reduced demand were countermanded by the realisation that the occupation of the modern MIE represented the 'new wave' in defence technology and procurement. Even in the relatively technologically-mundane AFV arena, some firms were convinced of the need to develop design capabilities to shore up their role in production. Daimler, for example, conceived the Ferret armoured scout car for the British Army in 1949, and proceeded to make 4,500 of them at Coventry between 1952 and 1971. Before retreating from the field in 1971, the firm oversaw the pre-production of a replacement vehicle, the Fox, which was put into quantity production by ROF Leeds. GKN, likewise, supplemented its earnings in civil markets with AFV work; although, unlike Daimler, it reaffirmed its military presence in the 1960s with sizeable production contracts for APCs. As a result of competitive tendering, it was allocated a contract to supply all the British Army's requirement for 1,048 MVC-80 APCs in 1985. These are to emanate from GKN-Sankey's Telford works.[7] Across the Atlantic, FMC gave itself over very largely to defence production after World War 2. In the early-1950s, for instance, it built M75 armoured personnel carriers in co-operation with International Harvester (ostensibly, yet another farm machinery and tractor manufacturer), and from 1960 retooled its San Jose, California, plant for series production of M113 APCs: an activity superseded by the production of M2/M3 Bradley infantry and cavalry fighting vehicles. Interestingly, FMC also retained an ordnance division in Minneapolis, Minnesota, for the purpose of making large-calibre artillery ammunition (e.g. 203mm shells for the US Navy). GM, too, persisted with AFV activity after the war. Its Cadillac division formulated the M56 Scorpion tank destroyer in 1948 and went on to build them during the Korean War. A Canadian subsidiary of GM, located in London, Ontario, currently makes under licence copies of the Swiss MOWAG Piranha APC for the US forces. Chrysler, of course, had emerged from World War 2 as a prime mover in MBT production, and it was to retain that position until, undermined by difficulties in its mainstream car activities, it was compelled to sell off its lucrative tank business to GD in 1982.

The more striking case of engineering enterprises entering into the world of modern MIEs, and achieving prime-contractor status into the bargain, is best illustrated through the postwar experiences of English Electric in the UK and GE in the USA. In point of fact, English Electric was a mongrel, formed in 1918 out of the amalgamation of the Coventry Ordnance Works, Dick Kerr & Company of Preston and the Phoenix Dynamo Manufacturing Company of Bradford. All of these predecessor firms had engaged in World War 1 aircraft production, but the consolidated successor determined to pare back such activity and had exited from the industry by 1926. The remorseless demands of rearmament, however, occasioned the re-entry of English Electric into aircraft manufacture in 1938. Three of the company's Preston plants were devoted to bomber production: an activity which was steadily built up through the war. After focusing on Handley Page-designed Hampden and Halifax bombers, the company had diverted its Samlesbury factory to the series production of de Havilland-designed Vampire jet fighters by the termination of hostilities. Bolstered by acquisition of a new facility at Warton and encouraged by the ease in which the Preston complex had mastered aircraft process technologies, the firm was keen to entertain further involvement in the industry at war's end. The resultant newly-assembled design team was soon to beget the estimable Canberra jet bomber in 1949. That event inaugurated a period of success for English Electric which, successively under the banners of the British Aircraft Corporation (from 1960) and BAe (from 1977), confirmed its place as the foremost UK warplane design and production complex. GE, on the other hand, chose to specialise in aeroengines.[8] A traditional MIE of long standing by virtue of its function as a supplier of steam turbines to the US Navy, the company had been persuaded to enter the experimental turbojet field during the war. Building on the UK Whittle W.1 prototype engine, GE's Lynn, Massachusetts, group devised engines for the first US jet aeroplane, the Bell P-59. By 1948, it had created the highly-successful J47 engine which was soon to be selected by the US military for mass production. Overwhelmed with orders, GE supplied J47s not only from Lynn but also from a GOCO plant near Cincinnati, Ohio, formerly operated by Wright. Now, restyled Evendale, that site is the

headquarters for GE's aeroengine activities. A follow-on turbojet, the J79, confirmed GE's eminence in
aeroengine design and production in the 1950s. Significant developments in turbofan and turboprop
engine technologies occurred in the 1960s and the
company entered the 1980s as the main competitor to
Pratt & Whitney in military aeroengine supply,
effectively usurping its New England neighbour in
the provision of powerplants for the F-14 and F-16
fighters.[9]

TECHNOLOGICAL PROGRESSION

It is logical to suppose that firms will advance
their product and process technologies in an incremental fashion, via learning economies, gaining
design confidence on the one hand and breadth of
production capability on the other. The classical
industrial life-cycle vision of enterprise evolution
is rooted in sequential thinking of this kind and
the record of the pioneer designers, starting up
their own firms in the aircraft industry in the
early part of the century, conforms to type. In the
military sphere, however, firms can circumvent this
gradualistic unfolding procedure by skimping on
design and choosing, instead, to plunge into production through manufacture of the designs of other
MIEs. As well as buying R & D and design expertise
off the shelf, so to speak, they also secure 'turn-
key' production technology, often with the key personnel needed to execute it. As a result, the firm
obtains relatively expeditious entry into the aircraft market. Once sufficient experience has been
acquired, the 'licence producer' may venture into
limited areas of design: an operation almost invariably tackled by way of marginal amendments to aeroplanes of the licensed-production variety currently
under construction. Design accomplishments accrue
from the culmination of such minor initiatives and,
eventually, the enterprise may branch out into
larger developmental projects in conjunction with a
more seasoned risk-sharing partner. At that point,
the MIE has transcended the limits of the 'dependent
co-producer' to adopt the guise of a 'co-operative
prime'. In due course, the enterprise will become
well-versed in design and production skills: so much
so, in fact, that it will be prepared to take on the
duties of a 'lead prime' and internalise the entire
spectrum of aircraft production from indigenous
design to product completion.
     One firm which stands testimony to the apparent

The Modern MIE

veracity of this alternative to the classic industrial life-cycle is Westland. The UK helicopter
enterprise commenced operations in World War 1, making aeroplanes to the designs of Airco, Shorts, Sopwith and Vickers. An element of upgrading occurred
when the Yeovil factory was tasked with converting
the Airco DH.9 to the DH.9A and then given contracts
to build almost 400 of them. While prototypes to the
company's own designs issued from the workshops in
the 1920s, it was only with the major re-design of
the DH.9A and its transformation into the Wapiti--
effectively shifting Westland from a 'dependent co-
producer' to a 'co-operative prime'--that the firm's
production success was assured. By the verge of
World War 2, Westland had emerged as a fully-fledged
designer in its own right, going on to produce the
Lysander, Whirlwind, Welkin and Wyvern. In 1947 the
company instituted what was to become a replay of
the process, purchasing licence-production rights
for helicopters from Sikorsky of the USA.[10] A decade
later, Westland (soon to embody other disparate UK
helicopter interests, especially the former Bristol
Aeroplane plant at Weston-super-Mare) was wholeheartedly committed to helicopter work, initially
producing identical copies of Sikorsky machines
before grafting indigenous design features on to
succeeding models. In the 1960s, the company entered
into a major co-operative agreement with Aérospatiale which was forthcoming with a family of jointly-
designed helicopters (the Puma, Gazelle and Lynx).
Perversely for the purity of the technological progression, however, Westland is manufacturing, at
present, an improved model of the Sikorsky Sea King,
the Model 30 of indigenous design (albeit, incorpo-
rating elements of the Lynx), the EH.101 which is
the outcome of joint design undertaken with Agusta,
and is fully prepared to assemble Sikorsky Black
Hawks involving scarcely any design changes.
Clearly, the technological sequence of development
is not inviolable. Indeed, some MIEs have displayed
little inclination to adhere to such a sequence:
rather, they simultaneously undertake two or more
facets of it, and have done so from the very outset
of their operations. Thus, as the contemporary
Westland example evinces, an aircraft manufacturer
may, at one and the same time, entertain the prospect of manufacturing under licence together with
manufacturing of in-house designs. More commonly,
firms can alternate between licensed-production and
manufacture of indigenous designs. Indisputably,
though, many contemporary 'lead primes' began as
'licence producers', using the preliminary

214

experience both to reinforce production know-how and escalate the accumulation of corporate resources. It is to those cases that we now address our comments.

Licence Producers

One of the world's principal combat aircraft producers, MD, is a relative newcomer in comparison with other modern MIEs in the AICs. The founder of its warplane wing, James Smith McDonnell, was an enlightened innovator whose company--McDonnell Aircraft--was incorporated in St. Louis, Missouri, in 1939. In spite of promising design initiatives, however, the budding McDonnell Aircraft was obliged to rely on government contracts for established designs, and ran a plant at Memphis as part and parcel of its war programme for manufacturing Fairchild AT-21 trainers. If nothing else, that endeavour provided the company with the breathing space to allow the genius of James McDonnell to fructify. His innovative FH-1 fighter appeared in 1945: the first of a long line of Navy and Air Force combat aircraft to be conceived and built at St. Louis. Today, the company produces the F-15 Eagle, the F/A-18 Hornet (embodying considerable design input from Northrop) and the AV-8B Harrier (capitalising on BAe technology) in its St. Louis complex and, as a result of merger with Douglas in 1967, maintains a second design and production facility in Long Beach, California, which is exercised in transport and trainer manufacturing.

A doyen of present-day licensed-production operations is MHI. As a general defence contractor involved in warship and AFV construction in addition to aerospace ventures, MHI became a 'lead prime' in aircraft activities prior to World War 2. Its postwar revival in the aircraft industry, however, has been of more modest proportions. The four plants constituting its Nagoya complex are currently immersed in the business of manufacturing MD's Eagle fighter as the F-15J. Consonant with the strategy of alternating licensed-production with production of indigenous designs, MHI made MD's earlier F-4EJ throughout the 1970s and the Lockheed F-104 in the 1960s but interspersed these programmes with production of its own T-2 advanced trainer and F-1 tactical fighter. To ensure on-going production activity, MHI possesses other aircraft licenses, including one for the Sikorsky S-61 Sea King helicopter (recall Westland). Indeed, since their rebirth in 1953, MHI and the other Japanese aircraft makers have supplied no less than 19 different US military aircraft under licence.[11] Harking even further back, however, one can ascertain that Japanese modern MIEs owed much to foreign designers during their formative years. MHI, for example, evolved after World War 1 on the strength of engineering designs furnished to it by Herbert Smith, formerly of Sopwith in the UK. Other design assistance was provided to it by Dr. Alexander Baumann, a representative of Junkers and other German interests. KHI, for its part, relied on technology transfer and design assistance forwarded by Dornier, especially in the person of Richard Vogt.[12] As of 1988, KHI was concurrently manufacturing the P-3C Orion under licence from Lockheed and, attesting to its innate design capability, the Kawasaki T-4 jet trainer.

Dependent Co-producers

As mentioned above, MD assumes the role of a global leader in combat aircraft production by keeping up several active and contemporaneous production lines. Among them, the T-45 stands out as an example of incremental improvement. The programme is aimed at furnishing the US Navy with a total pilot training package, and its aircraft element (about 85 per cent by value of the package) is provided by a modified BAe Hawk. Since the Hawk is earmarked for production at one of MD's Long Beach factories, the US firm is essentially functioning as a 'dependent co-producer' to the UK instigator of the aeroplane. While much of the basic airframe remains faithful to the original Hawk (and, in fact, is supplied in assemblies by BAe), the T-45 contains modifications--overseen by MD and Sperry--making the machine eligible for carrier operations. The T-45 is truly a contemporary example of an incremental improvement programme implemented by a prime contractor which was not responsible for the original design, but many firms have carried out such amendments over the years in prime-contractor programmes of their own making. For example, Canadair built the Argus maritime patrol aircraft for the Canadian forces in the late-1950s. Using the airframe of the Bristol Britannia turboprop airliner, Canadair substituted Wright turbo-compound engines into the Argus and 'militarised' the machine so as to enable it to undertake ECM and ASW missions. At the same time, the Montreal company had revamped the North American F-86 Sabre and was producing the fighter in quantity for use by NATO air forces. The revamping was centred, for the most part, on fitting a Canadian turbojet, the Avro Orenda, into the US-designed airframe in place of the GE J47.[13] Completion of Argus and Sabre contracts compelled Canadair to revert to 'licence producer' status in the 1960s with much of its workload deriving from the CF-104 (i.e. Lockheed F-104G) programme. KHI, too, found itself bracketed in the 'dependent co-producer' category in consequence of a maritime patrol aircraft project. After building 48 copies of the Lockheed P-2H Neptune from the late-1950s, it developed the P-2J and, after 1966, turned out one prototype and 82 production examples of this upgrade. The P-2J enjoyed updated avionics, turboprop power (GE T64) rather than piston engines, a redesigned undercarriage and rudder, and was somewhat longer than the original Lockheed design. Paradoxically, KHI has returned to the fold of 'licence producer' in supplying a replacement for the P-2J to Japan's Maritime Self-Defence Force. In this instance, the design donor is yet again Lockheed and the aeroplane in question is the P-3.[14]

Co-operative Primes

The category of 'co-operative prime' is regaining currency in the 1980s in the AICs, not so much because of the incidence of MIEs graduating from 'dependent co-producer' status, as it is owing to 'lead primes' abandoning go-it-alone strategies for international collaboration in weapons system design and production. The rationale for this tendency is well known and has been touched on in several places in this book. To all intents and purposes, it can be ascribed to the inexorable rise of aircraft development costs conspiring to force a response out of governments which takes the form of burden-sharing between them and, ergo, their respective national lead primes. The Eurofighter is the supreme exponent of this set of circumstances. Under the auspices of Munich-based Eurofighter GmbH, the project to develop a fighter for the 1990s will utilise the lessons learned in international management by the pathbreaking Panavia organisation, also housed in Munich. Unlike the three-nation Panavia scheme, though, the Eurofighter will embrace four nations, with Spain's CASA joining the original Panavia Tornado threesome of BAe, MBB and Aeritalia. As befits their standing as representatives of the major purchasers of the future fighter, BAe and MBB will each assume a 33 per cent share in Eurofighter whereas Aeritalia will take a 21 per cent stake and CASA will settle for a 13 per cent holding. An identical breakdown of shares obtains for the associated aeroengine consortium--Eurojet--which draws together Rolls-Royce, MTU, Fiat Aviazione and Spain's Sener.[15] No fewer than 800 Eurofighter aircraft are expected to ensue from the co-operative efforts of these MIEs. In order to bolster its fighter expertise, one of the partners in the consortium, MBB, is actively engaged with RI of the USA in the experimental X-31A Enhanced Fighter Manoeuvrability demonstrator programme.[16]

As an additional augury of the future, Japan's FS-X replacement for the F-1 tactical fighter is destined to be a thoroughly reworked version of GD's F-16. Both MHI and KHI are to be deeply involved as collaborators with the American firm (with MHI as prime contractor) in bringing the new support fighter to fruition. Expected to cost $1.1 billion to develop and entailing unit production costs amounting to $35 million, a modest 130 aircraft are to be built for entry into service in 1997.[17] Other contemporary co-operative ventures pulling together the assets of lead primes at government instigation include the AV-8B Harrier, the result of a pooling of the resources of BAe and MD (and Rolls-Royce in the engine domain) for the continued improvement of the original Hawker Siddeley warplane; the AM-X, a joint product of Aeritalia and Aermacchi (with, as we shall see, added input from Brazil); and a quartet of domestic American projects; namely, the Advanced Tactical Fighter, the Advanced Tactical Aircraft, the LHX and the V-22. The first consists of two consortia competing with each other to provide the US Air Force with its future fighter. One group, led by Lockheed and supported by Boeing and GD, has received a $691 million contract to produce a prototype designated the YF-22A. The other group is fronted by Northrop and embraces MD, and it also has received an equivalent contract to formulate a YF-23A prototype. By the same token, the two aeroengine companies are working on rival engines capable of insertion into either of the prototypes. Pratt & Whitney's entrant is designated as the YF119 whereas GE's goes under the YF120 marque. A fly-off competition is expected to determine the preferred airframe and engine in 1993.[18] The Navy's Advanced Tactical Aircraft, meanwhile, is supposed to have some commonality with the Air Force's future fighter, but will differ in its orientation to carrier-based operations. A buy of 750 Air Force machines and 550 Navy machines has been mooted. Not to be overlooked, the US Army is pushing its LHX programme to develop a new-generation anti-tank helicopter. Again, competing prototypes are required: in this case, the two teams in question consist of Sikorsky and Boeing on the one hand and Bell allied with MD on the other.[19] Finally, the V-22 Osprey is a tilt-rotor vertical lift aircraft under joint development by Bell and Boeing. They have received government advances of $1,714 million towards the total R & D cost of $2.5 billion. Anticipated orders from the Navy and Marine Corps sum to 1,213 machines, with the first deliveries due at the end of 1991.

Lead Primes

Fully-fledged design is the preserve of lead primes and, despite the aforementioned predilection of many of them to combine in collaborative programmes, several in the AICs persist in formulating and manufacturing combat aircraft, by and large, through using their own resources. MD has already been highlighted in these terms, but it is far from unique. Other American aircraft manufacturers manage to internalise significant warplane programmes and a listing of them follows below.[20]

Grumman.
This Long Island company discharges prime contractor duties to the Navy in several respects. In the attack aircraft realm it made the A-6E (six worth $310 million were procured in 1985) and has designed the follow-on A-6F. In the fighter sphere, Grumman sold 24 F-14A aeroplanes worth nearly $1 billion in 1985 and has its follow-on, the F-14D poised for production. For EW missions, it provides the Navy with EA-6B aircraft (six in 1985 worth $390 million) while for AEW it is forthcoming with the E-2C (six in 1985 worth $334 million).

GD.
The prime contractor for the F-16, the Fort Worth plant supplied 150 worth $2.8 billion to the Air Force in 1985 and was contracted to furnish an additional ten to the Navy worth $122 million.

MD.
For the record, 1985 program procurements for MD included $2.046 billion for 42 Air Force F-15 fighters, $2.417 billion for 84 Navy and Marine Corps F/A-18s and $696 million for 32 Marine Corps AV-8B tactical aircraft.

219

Northrop.
This company is pre-occupied with the 'stealth' Advanced Technology Bomber (now known as the B-2). In connection with this project, the firm is commissioning a new $100 million manufacturing plant at Perry, Georgia, which will employ 700 workers.[21]

Lockheed.
Equally immersed in 'stealth' technology, Lockheed reputedly has built numbers of F-19 fighters from its so-called 'Skunk Works' in California. Additionally, it was contracted in 1985 to supply the Navy with nine P-3C Orions worth $416 million.

Boeing.
Constituting the backbone of the US airborne EW effort, Boeing maintains a variety of AWACS and EW aircraft programmes including the E-3A, E-6 and E-8.

Of course, lead primes are not confined to the USA. In Europe, BAe is noteworthy for its innovative Sea Harrier which has resulted in a dedicated naval V/STOL fighter aircraft. Even more so, Dassault-Breguet stands out as a principal upholder of 'lead prime' status with all that is thereby entailed in terms of design independence. Such advocacy denied Dassault-Breguet a role in the Eurofighter consortium, and the firm has opted, instead, to develop its own future fighter, the Rafale. In the interim, it is supplying the French forces with the nuclear-capable Mirage 2000N. All told, France is expected to take 291 of all models of the Mirage 2000 and the company has accrued export orders for it from Abu Dhabi, Egypt, Greece, India and Peru. Equally insistent on its independence is Sweden's national champion, Saab-Scania. The prototype of that firm's new combat aircraft, the Gripen, flew in 1987 and at least 300 are required by the Swedish Air Force. Saab's independence is qualified, however, since the Gripen embodies substantial amounts of foreign technology.[22] For example, BAe designed much of the wing structure while other UK firms contributed to the landing gear (AP Precision Hydraulics), gearbox (Dowty Rotol), power generator (Lucas Aerospace), ejection seat (Martin-Baker) and radar (Ferranti). American involvement included the engine design (GE), wheels (Goodyear), flight control systems (Lear Siegler), generators (Sundstrand) and head-up display (GM Hughes). French and West German suppliers cannot be discounted either: the former's Microturbo is responsible for the auxiliary power unit whereas the latter's Mauser is contracted to supply the cannon armament.

NIC ENTRY

In light of the trend towards collaboration among AIC lead primes in the first place, and the quali-fied independence of those such as Saab-Scania which persist in go-it-alone programmes in the second place, it appears a little ironical that NIC air-craft firms often proclaim that they aspire to achieve 'lead prime' status. To a considerable extent, they are influenced in this desire by the Soviet example. There, they see a whole host of mod-ern MIEs functioning as state enterprises and seem-ingly effectively internalising the aircraft produc-tion process. India has utilised Soviet designs to provide its own state enterprise, HAL, with experi-ence of quantity production.[23] Nevertheless (and most significantly), it draws on AIC sources for design assistance, as is attested by Dassault-Bre-guet's input into the country's next-generation aer-oplane, the Light Combat Aircraft, and MBB's partic-ipation in India's new light helicopter project. China is the leading NIC exponent of lead primes. After the rupture with the USSR in 1960, the coun-try's burgeoning aircraft industry had all its Soviet aircraft production licences annulled. Left to its own devices, China resorted to 'reverse engi-neering', which is to say, it made unlicensed copies of Soviet aircraft. In 1964, for instance, it began production of a copy of the MiG-21 called the J-7 and in 1980 it flew a highly-modified MiG-23 known as the J-8.[24] In recent years, it has supplemented its Soviet-derived offerings by aircraft of a West-ern stamp. For example, the Nanchang enterprise is working on a new jet trainer, the L-8, which relies on a US Garrett TFE731 turbofan for its power-plant.[25] Similarly, the J-8-II update of the J-8 is to be fitted with US equipment, in this instance a fire control system supplied by Grumman Aerospace.[26] As far as exports are concerned, the Xian plant is producing the F-7M version of the J-7 and is relying on Western avionics to make it marketable. A new version of the A-5 attack aircraft, exported in quantity to Pakistan, is incumbent on avionics sup-plied by Aeritalia. These ventures notwithstanding, China continues small-scale manufacture of bombers and transports derived from Soviet designs and has complemented them with genuine licensed-production of Western helicopters (i.e. Aérospatiale's Dauphin and an improved version of the Super Frelon).

Moreover, it has blended older Soviet technologies into new products. The Harbin PS-5 amphibian, for example, appears to have the wing and turboprop engines of the An-12 transport combined with a tail section derived from the Be-6 flying boat and a newly-designed fuselage.

For reasons similar to those which obtained on the mainland, Taiwan has been severed of much of its foreign supply of weaponry. Consequently, it is frantically attempting to project state-owned AIDC into the ranks of the lead primes. The Taichung enterprise is already capable of producing AT-3 attack trainers (equipped with the same Garrett turbofan as the impending Chinese L-8) and promises to come up with the Indigenous Defence Fighter by the early-1990s. This latter is to utilise US engines (Garrett TFE1042) and avionics (Lear Siegler). The AIDC expects to put to good use the production facilities built up since the early-1970s in connection with the licensed-production of the Northrop F-5E/F.[27] Fellow NIC, South Korea, also professes serious aerospace ambitions. Under the aegis of the state airline, it also has undertaken licensed-production of F-5E/F aircraft. In addition, Korean Air manufactures the MD 500 light combat helicopter. The Samsung conglomerate acquired a licence from Bell-Helicopter-Textron to produce the model 412SP helicopter, and ground has been broken for a new plant at Sanchon. Samsung Precision Industries was first exposed to aerospace work in 1980 when it began overhauling the GE J85 engines used in the F-5s. Concurrently, another conglomerate, Daewoo Heavy Industries, has formed a joint-venture company with Sikorsky and is to make under licence the US firm's S-76 helicopter in South Korea. The country has a requirement for 120 new fighters and has appointed Samsung to act as prime contractor for determining the appropriate foreign design which would serve as the basis for indigenous production.[28] While offering great promise, these East Asian incursions into aircraft design and manufacture still fall considerably short of the level of skill attained in the AICs. However, other NICs have made even more material gains in modern MIE capabilities and they are deserving of special treatment.

MATURITY OF NIC DEPENDENT PRODUCERS

Both Israel and Brazil have progressed far along the path required to be taken in developing modern MIEs. Their endeavours, beginning in the 1950s, rested

initially on licensed production of designs originating in the AICs, but latterly shifted to the promoting of genuine indigenous lead primes. Rapidity
informs much of the development in both cases. So
far as Israel is concerned, state enterprise IAI
began as an aircraft overhaul facility in the early
1950s (when it was known as Bedek), although it had
graduated to licensed manufacture of French Potez
Magister jet trainers by the end of the decade. IAI
became a force to be reckoned with after the 1967
'Six Day' War. French embargoes on supplies of Das-
sault Mirage fighters spurred IAI into production of
Nesher and Kfir derivatives of that aircraft; a ploy
aided by the provision of GE J79 turbojets for the
latter aeroplane in place of the original SNECMA
powerplants. Thus, for reasons similar to those
which obtained in China after 1960, Israel was compelled to resort to 'reverse engineering' of foreign
designs as a means of sustaining the defence-industrial base. Consequently, in ten years IAI expanded
from a workforce of 4,000 to 23,000 and entered the
1980s with high hopes of formulating (with US technical and financial assistance) a new-generation
fighter, the Lavi. Subsequent withdrawal of US aid
commitments, rising development costs and the possibility of Lavi expenditures depriving the military
of other much-needed defence items, led the Israeli
Government to formally abrogate the Lavi programme
on 30 August 1987. Cancellation placed up to 4,000
IAI jobs in jeopardy and not only dashes Israel's
hope of retaining IAI as a lead prime in the warplane sphere, but casts doubt on Bet Shemesh
Engines, the country's sole supplier of aeroengines.[29]
    Less ambitious but scarcely less remarkable has
been the emergence of Brazilian modern MIEs. In the
field of AFVs, private firm ENGESA built on the success of armoured car exports to achieve proficiency
in MBT design and production. While manufacturing
its own armour plate, ENGESA manages to tap Brazil's
extensive automotive industry to purchase most of
its AFV components from production lines geared to
civilian items.[30] Located at Sao José dos Campos,
ENGESA shares an arms industrial complex with the
Avibras private MIE and the EMBRAER part-state
enterprise. In combination, they are the result of
the Brazilian state, buoyed by desires of regional
power projection, both assuming direct involvement
in military industry and enlisting the support of
private capital in the formation of a defence
enclave. Avibras, for instance, ranks as the country's 22nd largest exporter with 1985 exports

reaching $171 million. Employment totals 6,600 (mid-1987), including 500 engineers; a marked increase on the total employment figure of 250 recorded in 1978. The company's leading product line and export earner is the Astros 2 rocket artillery saturation system, which consists of a family of mobile rockets of different calibres and operational ranges of up to 60 km, and includes a universal multiple launcher, an optional fire control unit and an ammunition supply unit.[31] To market a complete weapons package, Avibras established Tectran to manufacture the 10-ton armoured truck which will carry Astros 2 systems. In addition, the firm produces a variety of aerospace weapons systems and components, not least of which are the Barracuda coastal defence missile and the SS-300 battlefield missile. Its future strategy is centred on the development of a new series of anti-ship missiles to challenge Aérospatiale's Exocet. Avibras is hoping to extend its technological capabilities by undertaking development of inertial guidance platforms for anti-ship missiles on the one hand, and by developing an advanced version of the Swiss Contraves air defence system on the other. In respect of the latter, the Army has signed a $100 million development contract that includes procurement of the first 13 units.

Testimony to the strength of state-controlled EMBRAER is the fact that it can obtain capital from international sources notwithstanding Brazil's economic problems. Indeed, EMBRAER's performance has been exceptional for what is still a relatively nascent aerospace venture. For example, its sales have doubled (to $378 million) between 1982 and 1986, while employment has climbed from 6,732 to 8,592. The firm's product line revolves round the Brasilia commuterliner and the Tucano turboprop-powered military trainer, the latter being the subject of orders and options for 586 aircraft from eight nations (including the UK, where it is built under licence by Shorts). The Brasilia is the successor to the Bandeirante, the landmark project which launched the company. No less than 470 of the earlier aeroplanes were built, many being sold to US regional airlines. Yet, the Bandeirante's success rested, in no small measure, on the backing of the Brazilian military. EMBRAER itself was started in 1969 with the object of manufacturing the Bandeirante, a machine developed by the Centro Tecnico Aero-Espacial (a government research centre) expressly to meet an Air Force requirement for a light transport. A key policy decision in the success of the Bandeirante programme was the Air Force's procurement of Italian Aermacchi

MB-326 jet trainers (restyled Xavantes in Brazil) which were allocated to EMBRAER for licensed-production. Significantly, given its limited experience of the complexities of aircraft manufacture, the Xavante endeavour allowed EMBRAER to acquire the know-how for establishing series production, including its amassing of knowledge in such areas as tracing technology, technical documentation, production planning, quality control, and tooling.[32] In short order, these techniques were transferred to the Bandeirante programme. Furthermore, EMBRAER was enabled to advance down the learning-curve by dint of the Air Force's crucial launch order for 80 Bandeirantes. Thus, through its procurement decisions, the Air Force has been a vital agency in the fostering of an indigenous aircraft industry and its presence has been felt not only as EMBRAER's major domestic market, but perhaps more importantly, as a result of its willingness to act as a 'co-operative' customer. In other words, beyond simply buying aeroplanes, the Air Force injected management support, provided investment for product development, and allowed EMBRAER to capitalise on market opportunities by adapting its procurement programmes to the company's more general commercial interests. EMBRAER's current co-operation, at government connivance, with Aeritalia and Aermacchi in the AM-X tactical fighter programme is further proof of the critical importance of Air Force underwriting of production activity.

While illuminating in their own right, the experience of the two countries is salutary for others. The Israeli and Brazilian determination to create modern MIEs with far-reaching technical and production attributes is crucially significant for what it portends. In short, their efforts foreshadow the desire of all NICs to foment modern MIEs that go beyond the dependent producer model. The Israeli experience attests to the possibility of attaining that goal—albeit with considerable help from AIC bankers and technology donors, and for what may be only a brief span. Conversely, the Brazilian case intimates that the status of co-operative prime may be more readily achievable and, therefore, a more realistic target for NIC licence-producers to aim for. At any rate, there is no doubt that weapons procurement is being increasingly meshed with military-industrialisation ambitions both in the NICs and AICs. Furthermore, the offsets arrangement is a common instrument deployed by governments in their attempts to bring such goals to fruition. It is to deliberations of the integration of offsets with the moulding of modern MIEs that we now turn.

OFFSETS

Relationships between lead primes and dependent producers among the global set of modern MIEs cannot be
divorced from the 'internationalisation' of military
-industrial production mentioned at the outset of
this book. In particular, the lead prime and its
national government feel impelled to export and so
contain costs of weapons by recovering a portion of
R & D outlays, extending production runs to realise
scale and learning economies, and keeping production
lines active when domestic procurement programmes
are unable to sustain capacity utilisation. While
all these objectives are difficult to achieve,
exports at least offer the prospect of conserving
the defence-industrial base, have a positive impact
on the trade balance and result in profits for the
enterprise. Yet, the striking contemporary reality
is that compensatory arrangements are increasingly
necessary to secure export sales. These arrangements
are designed to 'offset' the cost of procuring
expensive weapons on the part of the buyer through
the recovery of hard currency, employment creation
in the buyer country, support of the buyer's industrial base and, significantly, technology transfer
from seller to buyer. Thus, offsets allow the purchasing government to realise multiple political and
economic goals while, simultaneously, enabling the
supplier to find an outlet for its wares. Not only
are they often conceived as an instrument of nation
building by the buyer, as they are claimed to reduce
dependence on foreign weapons in the long term, but
offsets are espoused because of their presumed
employment and multiplier benefits in the purchasing
country.[33]
    Concomitantly, and the focus of this section,
these agreements are increasingly being used as
regional-industrial policy, be it in the context of
developing indigenous military-industrial capability, targeting industries and firms, or enhancing
the competitiveness of local industry generally. A
central objective in this regard is to use market
leverage to establish conduits for transfer of technology. For example, the then Australian Defence
Minister, D. J. Killen, expressly emphasised the
technology transfer aspect of his country's new
fighter offset strategy of the late-1970s, stating:
'Investment is needed to introduce into our industries the techniques of these advanced aircraft'.[34]

In the event, selection of the MD F-18 was tied to offsets amounting to 40 per cent of the $2.75 billion deal (since revised to A$4 billion). Under the terms of the so-called Australian Industry Participation (AIP) programme, F-18 airframes are assembled in Australia by GAF while the fighter's GE F404 engines are assembled by CAC. More importantly, the AIP programme targets aerospace technology transfer as a vital component of the entire offset strategy. Through it, Australian industry gains access to such process technologies as titanium hot forming and chemical milling, aluminium no-draft precision forging, five-axis machining, new plating processes, manufacturing technology related to graphite epoxy composite structures, and hot and cold bonding techniques.[35] Acquisition of technology of this kind will enable Australian industry to provide life-cycle support for the F-18 and serve as a springboard to long-term competitiveness. The latter focus continues to inform Australian policy. An official pronouncement in 1986 vindicated offset policy on the strength of its 'competitive technology focus' which, in practice, meant that 'overseas suppliers to the Commonwealth of goods and services will be required to bring direct work or technology to Australia...with the aim of bringing to Australian industry advanced technology, skills and capabilities'.[36]

In detail, there are five basic types of industrial and commercial compensation practices associated with these agreements; namely, co-production, licensed production, subcontractor production, technology transfer and foreign investment.[37] The first is a government-to-government arrangement permitting the purchaser to acquire the technical information needed to manufacture a defence article. When refined to conform to the terminology introduced in Chapter 3, co-production entails a lead prime acting as seller and a dependent co-producer serving as buyer in two different countries. What is more, it assumes that the dependent co-producer has sufficient autonomous capability to modify or redesign the original weapon system. Licensed production, however, is confined to the transfer of technical information under direct commercial arrangements between the lead prime and a foreign producer. No significant technological embodiment is expected to be forthcoming from the licence-producer MIEs. For its part, subcontractor production covers the manufacture of a part or subsystem by a foreign MIE which is subsequently incorporated into the system assembled by the lead prime. The subcontract does

227

not necessarily involve technology transfer between the lead prime and the subcontractor. In contradistinction, technology transfer invariably leads to the sharing of the prime contractor's proprietary knowledge with a foreign dependent producer. Finally, in our context, foreign investment refers to capital injections tied to an offset, frequently made manifest through the provision of specialist defence plant. Alternatively, offsets may be visualised from the standpoint of the distinction made between direct and indirect agreements. The former include any business related directly to the product being sold: thus, if the article is an aircraft, the direct offsets embrace aircraft parts manufacture, aerospace process technology transfers, direct investment in aircraft plant and the like. Conversely, indirect offsets include all business encompassed in the deal which does not spring directly from the defence article in question. For example, the lead prime may be asked to facilitate the marketing of the buyer country's agricultural produce in order to counter the currency outflows incident to the weapons deal. In practice, military-related sales agreements usually combine a number of offset types.

Particularly relevant to the link between defence procurement and regional-industrial policy is the implication for modern MIEs. In a US Treasury Department survey covering the period 1975–81, it was found that fully $8.94 billion out of the $9.55 billion incurred in 130 offset arrangements were tied to the sale of aircraft and related equipment.[38] Similarly, over the 1980–4 period, aircraft figured as the main constituent of offset arrangements. That the USA holds important technological advantages in aircraft manufacture is not lost on buyer governments, as they are able to use their purchasing leverage to negotiate transfer of process and product technologies and hence augment the capabilities of their own modern MIEs. Direct offsets tend to be more important than indirect offsets when measured in terms of value of commitments, with co-production and subcontractor production being the principal types demanded. The predominance of direct offsets, of course, confirms the desire of buyer governments to develop a modicum of defence industry as part of their industrial development thrust. It is now opportune to detail a few case studies.

Boeing and Saudi Arabia

Saudi Arabia has grasped the opportunity afforded by offsets to set up a defence industry from scratch. It has obliged, for instance, ENGESA of Brazil to agree in principle to the establishment of a factory at Jeddah for the manufacture of Osorio MBTs.[39] More far reaching, however, is the offset deal concluded as a result of the sale of Boeing air defence systems, E-3A AWACS and KE-3A aerial tankers to the country. Together, these ventures are worth something in the range of $4 billion. Known as the 'Peace Shield' programme, the offset deal arranged with US contractors requires that some 35 per cent of the hardware costs of air defence upgrading must be defrayed through offsets aimed at industrial investment in Saudi Arabia. Managed by the Boeing Industrial Technology Group in which Boeing is joined by Westinghouse Electric and ITT's Federal Electric Corporation (among others), the offset programme consists of two five-year phases. In the first, onus is placed on four major projects: an aerospace maintenance, repair and service centre (under the auspices of Boeing and GE); an advanced electronics centre (for which ITT is responsible); a computer systems servicing company; and a digital telecommunications production and servicing facility. The second phase calls for five additional joint ventures; namely, a Boeing-run helicopter plant ultimately dedicated to the production of Model 360 (or similar) aircraft, a Westinghouse-led power engineering centre, a Boeing-sponsored applied technology centre, an ITT-inspired biotechnology venture and, lastly, technology transfer to enable the manufacture of medical products. Geographically, the aerospace-related projects are collocated at the El Kharj Air Force Base, hard by a small-arms manufacturing arsenal, in order that they may interact to foster agglomeration economies. Significantly, the very comprehensiveness of the 'Peace Shield' programme has been held culpable for difficulties faced by other Western defence contractors which have, in duty bound, to arrange offsets in Saudi Arabia. For example, the UK firms involved in the £5 billion contract to sell Panavia Tornado, BAe Hawk and Pilatus PC-9 aircraft to the country complain that the one-third of the contract value which is to be defrayed through offsets cannot be easily achieved owing, in part, to the prior identification by Boeing of the most suitable opportunities.[40]

GD and Turkey

Turkey had fleeting experience of aircraft manufacture in the post-World War 2 period, but had to
await an offset programme struck with GD in order to
acquire a fully-fledged, 1980s-vintage aircraft
enterprise.[41] Under the 'Peace Onyx' programme
signed in May 1984, Turkey agreed to buy 160 GD
F-16s for an estimated price of $4.5 billion.[42]
Prime contractor, GD, is scheduled to provide $150
million in direct offsets and $1.27 billion in indi-
rect offsets, with obligations spread over ten
years. The major F-16 subcontractors are also
involved in the package. Thus, GE is setting up
licensed production of F110 turbofans at the new
TUSAS Engine Industries factory in Eskisehir while
Westinghouse is to manufacture APG-68 radars courtesy of its Turkish subsidiary, WESSA. Of paramount
importance, though, is the aircraft enterprise,
TUSAS Aerospace Industries, a joint venture of the
government (49 per cent), GD (42 per cent), General
Electric Technical Services (seven percent) and various service organisations. Its Murted complex is
expected to encompass greater than one million
square feet of floor space and is the beneficiary of
process technology transferred from GD, including
sheet metal forming, structural bonding, and five-
axis machining. In view of the need to gradually
build up production expertise, the Turkish F-16 programme will advance through a phased process, beginning with assembly of knock-down kits supplied by
GD. Subsequently, TUSAS will progressively expand
its manufacturing role as learning-curve economies
are realised until the condition is reached where
local content embraces 50 per cent of airframe
input. With increasing production competence, TUSAS
will manufacture its own aft and centre fuselage
sections, wingboxes, and the tools required for
these assemblies. Eventually, once indigenous production is firmly established, the Turkish MIE will
supply F-16 components directly to the lead prime.
In so far as indirect offsets are concerned, GD is
pursuing two programme areas. Firstly, it is committed to marketing Turkish souvenirs, videocassettes
and textiles in the USA while, secondly, within Turkey itself, it is developing the energy sector
through construction of a windmill manufacturing
plant, solar energy systems and a 960MW thermal
power plant.

Krauss-Maffei and Canada

Canada pioneered offset deals in the 1970s, partly to bailout its airframe industry, partly to boost regional development and, in part, to promote a patchy defence industry.[43] Supposedly inspired by the last consideration was the deal struck in 1976 between the Canadian Government and Krauss-Maffei for the purchase of 128 Leopard 1 MBTs worth an estimated C$187 million (later revised to C$236 million).[44] The government made the order conditional on Krauss-Maffei providing compensating industrial benefits to Canada to the extent of 40 per cent of the total contract, and the German MIE was given ten years to fulfil the obligation. As envisaged, these benefits were to come primarily in the form of industrial investment and counterpurchase of sophisticated goods. By oversight, and much to the chagrin of Canada, the tank deal was left open-ended with the effect that virtually anything in the way of manufactured products qualified as an offset credit. Consequently, the bulk of the offset work—amounting to C$81 million—was related to castings for manufactured goods or machinery, with little materialising in the way of purchases of advanced manufactures from Canadian industry. In fact, about nine per cent (C$8.4 million) of the offsets derived from purchases of plywood and airline tickets! The remainder was accounted for by purchases of rapeseed oil and contracts lodged—with Canadian Government concurrence—in the Netherlands. To all intents and purposes, only the Montreal precision-casting firm of Vestshell became involved in Leopard tank production. Thus, in as much as the offset agreement was visualised as an instrument for enhancing the historically-weak Canadian manufacturing sector, the Leopard programme did not deliver the intended benefits. Moreover, Ontario and Quebec firms captured 75 per cent of offset benefits—a clear violation of the government's implicit regional policy objective. This example of offsets, then, emphatically demonstrates that the procedure sometimes falls short as an instrument of technology transfer in general, let alone as a means for augmenting modern MIE capabilities in weapons-importing countries.

COMPENSATORY TRADE: A VIABLE STRATEGY?

In light of the above case, the efficacy of offsets in fostering modern MIEs in countries currently devoid of them is a matter subject to serious misgivings. In the first place, the imposition of offset conditions on arms sales contracts is not costless. Additional transaction costs, transfer of technology costs, duplication of R & D and capital investment, not to speak of unrealised production economies, all combine to push up the procurement cost of weapons systems. It has been calculated, for example, that the four nations involved in the European co-production of F-16 fighters incurred a 34 per cent cost penalty relative to an off-the-shelf price.[45] In the Japanese F-15 co-production programme, the cost per copy of MHI-supplied aircraft is $50 million as against a fly-away-price of half that sum for aircraft directly bought from MD.[46] All told, it has been estimated that Japan will need to find $1.8 billion more as a result of its insistence on domestic production of 90 F-15s. For their part, Canadian Aurora (i.e. Lockheed P-3 Orion), MD F-18 and patrol frigate offset programmes incurred administration and direct programme costs of at least C$375 million. In truth, those projects suggested that a 5-10 per cent cost premium is required to secure any industrial benefits.[47] Secondly, despite negotiation of offset obligations often exceeding the actual sales contract values, fulfillment of them has frequently proved elusive. GD, in the case alluded to above, was unable to fulfil its indirect offsets obligations to Turkey and had to settle for just over half of the scheduled offset credits in 1986.

Governments may be disposed to accept the higher costs associated with offsets if regional-industrial policy objectives are effectively realised. The reasoning here basically parallels many of the arguments favouring state intervention in general. To be specific, it is often maintained that if medium to long-term benefits attributable to offset agreements are deemed sufficient, that is, if social rates of return exceed opportunity costs of the additional expenditures incurred, then a sound case exists for making use of offsets as instruments of industrial policy. More important from a pragmatic viewpoint is the question of whether countries are able to utilise offsets for the purpose of diversifying exports and augmenting domestic industry, especially defence production. Canadian experience suggests that the potential to use offsets in this way is subject to limitations: in effect, the outcomes for domestic industry are likely to lead to built-to-print manufacturing capabilities at best, with little improvement in technological proficiencies. This finding is certainly not repudiated by the Taiwan experience where the Northrop-AIDC F-5E/F programme has fallen short of expectations. Originally intended to provide start-up AIDC with a domestic combat aircraft design faculty, the programme has scarcely progressed beyond the licence manufacturing stage. While the Taiwanese have steadily increased their participation in the project from nominal assembly to component fabrication, little has been achieved in critical areas of advanced manufacture, such as avionics, pneumatics or aeroengine fabrication. Northrop itself feels that no R & D, production design or qualification technology has been transferred as a result of the programme.[48] Moreover, the Taiwan example underscores the political constraints on using offsets as a basis of industrial development. US Congressional restrictions, arising from the sensitivities of US relations with both the People's Republic of China and the Republic of China (Taiwan), have continued to limit the flow of military technology to Taiwan. These restrictions have been particularly acute in the aeroengine field and have hitherto frustrated Taiwan's attempts to promote a well-rounded modern MIE infrastructure. Therefore, it is patently evident that offset agreements are susceptible to restrictions on technology flows imposed by both the vendor--the lead prime--in defence of its proprietary interests and by the vendor's government in the name of its political interests. Consequently, offsets are not in themselves assured means of acquiring an 'instant' defence industry. The more complex the technological basis in the sequence from traditional to emergent defence technologies, the less likelihood that offsets will work to effect a free-standing military-production capability. Considerable evidence, in fact, points to the inadequacy of offsets when they are vaulted to the position of serving as a platform for creating effective modern MIEs out of a scanty pre-existing defence-industrial base. Their worth is even more questionable in the context of high-technology defence electronics: a subject to which we devote the next chapter.

NOTES AND REFERENCES

1. Refer to D. Todd and J. Simpson, The world aircraft industry, (Croom Helm, London, 1986) and D. Todd and R. D. Humble, World aerospace: a statistical handbook, (Croom Helm, London, 1987).

2. As recorded in Bill Gunston, The plane makers, (Basinghall Books, London, 1980), p.116.

3. See Jane's all the world's aircraft 1919, (Arco Publishing, New York, 1969) for these and subsequent examples.

4. E. R. Zilbert, Albert Speer and the Nazi Ministry of Arms, (Associated University Presses, East Brunswick, NJ, 1981), pp.191-2.

5. W. L. Naumann, The story of Caterpillar Tractor Company, (The Newcomen Society, New York, 1977), p.12.

6. M. Donne, Leader of the skies: Rolls-Royce, (Frederick Muller, London, 1981), pp.46-52.

7. See commentary in The Engineer, 20 and 27 June 1985, p.14 and p.12 respectively.

8. In this, it was following the example set by engineering companies during World War 1 which, when wishing to enter the aircraft business, decided to undertake engine production. As well as Rolls-Royce, they included Allis-Chalmers, Gnome and Hitachi. Automotive firms, too, frequently opted for aeroengine involvement. Alfa Romeo, Allison, Alvis, Benz, Chrysler, Isotta-Fraschini, Lycoming, Packard and Renault all conform to this mould.

9. Bill Gunston, World encyclopaedia of aero engines, (Patrick Stephens, Wellingborough, 1986), pp.59-69. Note, GE also entered gun manufacture at the end of World War 2 as part of its thrust into aircraft markets. The first marketable product of its Burlington, Vermont, Aircraft Equipment Division was the M61A1 cannon, mounted in 1954 on the Lockheed F-104A fighter.

10. Sikorsky, representing its UTC parent, has been a minority shareholder in Westland since 1986.

11. See the article in Interavia, September 1986, pp.985-6.

12. E. Sekigawa, Pictorial history of Japanese military aviation, (Ian Allan, London, 1974), p.21.

13. K. M. Molson and H. A. Taylor, Canadian aircraft since 1909, (Putnam, London, 1982), pp.408-9.

14. Interestingly, Canada's Argus replacement was also the P-3. Unlike Japan, however, Canada preferred to buy the aircraft (renamed Aurora) directly from Lockheed.

15. Note Aviation Week & Space Technology, 9 June 1986, p.28 and Flight International, 13 September 1986, p.8.

16. Details are to be found in Aviation Week & Space Technology, 30 November 1987, pp.26-7.

17. As announced in Aviation Week & Space Technology, 26 October 1987, p.22.

18. See Flight International, 1 August 1987, p.34.

19. Refer to article on anti-tank helicopters in _Interavia_, September 1987, pp.921-6.

20. Abstracted from AIAA, _Aerospace facts and figures 1986-1987_, (Aerospace Industries Association of America, Washington, DC, 1986).

21. Noted in _Aviation Week & Space Technology_, 29 September 1986, p.26.

22. For an appraisal of the Gripen, see _Interavia_, August 1986, pp.867-70.

23. HAL manufactured in excess of 500 MiG-21 fighters and is currently licence-producing the MiG-27M.

24. V. Nemecek, _The history of Soviet aircraft from 1918_, (Willow Books, London, 1986), p.107 and p.110. The J-8 was inspired by a MiG-23 passed on to China by Egypt.

25. See _Flight International_, 20 June 1987, p.3.

26. Refer to _Aviation Week & Space Technology_, 15 August 1987, p.248.

27. The topic is followed up in _Aviation Week & Space Technology_, issues of 31 March 1986, p.31; 21 April 1986, p.77; and 6 July 1987, p.15.

28. Note the article in _Flight International_, 21 March 1987, pp.36-9.

29. See _Flight International_, 12 September 1987, p.8. Interestingly, the Army's tank arsenal—builder of the indigenous Merkava MBT, another side-effect of 1967 embargoes—may receive a new lease of life, rendered possible as a result of procurement programmes restored by the Lavi cancellation.

30. Note P. Lock, 'Brazil: arms for export' in M. Brzoska and T. Ohlson (eds.), _Arms production in the Third World_, (Taylor & Francis, London, 1986), pp.79-104; W. Perry and J. C. Weiss, 'Brazil' in J. E. Katz (ed.), _The implications of Third World military industrialization_, (Lexington Books, Lexington, Mass., 1986), pp.103-117; and C. Vanhecke, 'Brazil breaks into arms sales in a big way', _Manchester Guardian Weekly_, 6 April 1986, p.12.

31. As specified in _Aviation Week & Space Technology_, 17 August 1987, p.49.

32. Refer to C. J. Dahlman, 'Foreign technology and indigenous technological capability in Brazil' in M. Fransman and K. King (eds.), _Technological capability in the Third World_, (Macmillan, London, 1984), pp.317-34 and R. Ramamurti, 'High technology exports by state enterprises in LDCs: the Brazilian aircraft industry', _The Developing Economies_, vol. 33 (1985), pp.254-80.

33. S. G. Neuman, 'Countertrade, barter, and countertrade: offsets in the international arms

market', <u>Orbis</u>, vol. 29 (1985), pp.183-213.

34. Cited in <u>Aviation Week & Space Technology</u>, 10 December 1979, p.54.

35. See, for example, <u>Aviation Week & Space Technology</u>, 26 October 1981, p.23; <u>Flight International</u>, 16 March 1985, p.9; and <u>Interavia</u>, June 1987, pp.613-8.

36. S. M. Rubin, <u>The business manager's guide to barter, offset and countertrade</u>, EIU Special Report, no. 243 (Economist Intelligence Unit, London, 1986), p.51. The 1987 submarine deal struck by the Australian Government with Sweden's Kockums, necessitating the building of hulls at a purpose-built facility in Port Adelaide, is the most recent manifestation of this outlook. Note <u>Navy International</u>, vol. 92 (1987), pp.421-2.

37. Identified by the US International Trade Commission in its report, <u>Assessment of the effects of barter and countertrade transactions on US industries</u>, (USITC Publication 1766, October 1985).

38. Contained within the 'Report of a survey on offset and coproduction requirements', dated 24 May 1983. Along the same lines, see 'Trade offsets in Foreign Military Sales', GAO/NSIAD-84-102, dated 13 April 1984.

39. Reported in <u>Jane's Defence Weekly</u>, 8 February 1986, p.178. The deal has yet (end-1987) to be consummated, however, and the Saudi Government continues to evaluate a number of foreign MBT designs.

40. See <u>The Engineer</u>, 29 January 1987, p.6.

41. The Turkish military aircraft factory at Kayseri ceased aircraft manufacture in 1940. The Turkish Air League's Aircraft Factory at Etimesgut continued, under the MKEK label, to produce light aircraft into the 1950s.

42. Background information can be acquired from <u>Aviation Week & Space Technology</u>, issues of 14 May 1984, p.26; 6 April 1987, pp.70-1; and 15 June 1987, p.70. The topic is also scrutinised in <u>International Defense Review</u>, July 1987, pp.954-5.

43. For details of the first two reasons, see D. Todd and J. Simpson, 'Aerospace, the state and the regions: a Canadian perspective', <u>Political Geography Quarterly</u>, vol. 4 (1985), pp.111-30.

44. As recounted in the <u>Financial Post</u>, 27 November 1982, p.16 and the <u>Winnipeg Free Press</u>, 4 September 1987, p.13.

45. M. Rich et al, <u>Multinational coproduction of military aerospace systems</u>, (Rand, Santa Monica, Calif., 1981), p.viii.

46. See the <u>Financial Post</u>, 28 May 1983, p.S8 and Executive Office of the President, Office of

Science and Technology Policy, <u>Aeronautics research and technology policy</u>, (Washington, DC, November 1982).

47. Task Force on Program Review, Study Team Report, no. 8, 'Government procurement: spending smarter', Supply and Services Canada, Ottawa, March 1986, Appendix B.

48. Dealt with in J. E. Nolan, <u>Military industry in Taiwan and South Korea</u>, (Macmillan, London, 1986).

Chapter Six

THE EMERGENT MIE

It was announced at the end of 1986 that the Royal
Dockyard at Rosyth was to be turned over to commercial management for a seven year period extending
from April 1987 as part of the UK's plan to privatise state-owned MIEs. The consortium earmarked to
run the yard is Babcock-Thorn, a combination of
mechanical engineering and--courtesy of Thorn EMI--
electronics engineering interests. In the same
vein, albeit two years earlier, the first of the BS
specialist warship yards to be privatised, Yarrow of
Glasgow, was snapped up by electrical and electronics engineering giant GEC for £34 million. Both
these instances of traditional MIEs succumbing to
the control of electronics concerns highlight, in a
nutshell, the modern trend in defence production of
the diminishing importance of weapons platform producers in relation to the rising importance of the
makers of the systems contained within the platforms. In a word, the systems makers have overtaken
the platform producers in the pecking order of
defence manufacturers. Nor is the unsettling process
confined to traditional MIEs since the domain of the
modern MIE, aircraft, is equally susceptible to
encroachment by systems suppliers: one US systems
maker, Raytheon, has its own airframe subsidiary
(Beech) just as another, Litton, has its own warship-building subsidiary (Ingalls). The importance
of systems manufacturers in the aircraft industry
has grown in direct correspondence with the increasing stake owing to systems in aircraft production:
aviation electronics (i.e. avionics) which were
virtually nonexistent prior to the introduction of
airborne radars in World War 2 now make up as much
as half of the production cost of a combat aircraft.[1] Similarly, current-generation AFVs are obso-
lescent unless bedecked with a sophisticated systems
fit. The M1 tank, for example, requires a ballistic

computer (supplied by Computing Devices Company of
Ottawa, Ontario), a laser rangefinder and thermal
imaging system (from GM's Hughes Aircraft subsidiary), and a line-of-sight data link (obtained from
Singer's Kearfott Division). Thus, regardless of
platform type--ship, aircraft or MBT--the accent in
complexity and cost is shifting towards the systems
aspect of the weapon system and, therein, is compelling MIEs to refocus their attentions on systems
rather than platforms. In other words, systems makers are the 'emergent' MIEs, the more so as the systems revolve round a core expertise rooted in electronics technology which affords them advantages in
a plethora of products ranging from fire control
computers to the welter of sensors subsumed under
the EW banner. Two prominent defence programmes
undoubtedly dependent on resourcefulness in electronics are proffered forthwith in order to give
substance to the claim that systems prevail over
platforms in much of contemporary defence production
activity in the AICs.

The first, prosaically named Aegis, is a naval
weapons system: indeed, it is specifically a sea-
borne surface-to-air weapon system developed for the
US Navy.[2] Designed for installation aboard two
classes of warship built with the express purpose of
accommodating it, Aegis aims to provide the fleet
with the ability to deter anti-ship missiles. The
two ship classes in question, the 'Ticonderoga'
cruisers and the 'Arleigh Burke' destroyers, are the
principal platforms required for the US Navy's surface fleet through the 1980s and 1990s barring the
few additions to the nuclear-powered aircraft carrier force. Aegis, in fact, is a composite of several sub-systems; namely, the Standard SM-2 missile
along with its launcher, fire control and weapon
control systems; the ship's multi-function phased-
array radar and its accompanying computer control,
not to overlook the vessel's command and control
system as well as its operational readiness test
system. All these electronics sub-systems of Aegis
gear the ship and its missiles to the main function
of area air defence. The missile launcher has a
digital interface with the weapon control computer
which, in turn, accepts and then analyses tracking
information so as to engage the fire control system,
that is, a radar used to illuminate the target.
Surveillance and identification of would-be targets
is the lot of the multi-function phased-array radar
and this equipment depends on four computers which,
in conjunction, constitute the ship's command and
control system. The responsibility of RCA (now an

affiliate of GE) as prime contractor, AEGIS has been under development since 1964, with the first fully-equipped vessel, the 'Ticonderoga', becoming operational in January 1983. No fewer than three RCA divisions are involved in the programme and they are joined by more than 600 additional suppliers, the most important of which are Raytheon for the radar and transmitter, Sperry Univac (recently acquired by Burroughs) for computer hardware, Computer Sciences Corporation for computer software, FMC's Northern Ordnance Division for the missile launcher and GD's Pomona Division for the missile itself.

The second example is the Nimrod AEW, the ill-fated competitor to the E-3A AWACS of Boeing and Westinghouse for the UK's airborne early-warning defence system. Commissioned in 1977 under the wing of the Marconi Avionics subsidiary of GEC (later transferred to GEC Avionics), the project entailed the conversion of BAe Nimrod maritime patrol aircraft into a force of eleven AEW Mark 3 Nimrods for a bolstered UK air defence system, complementing that of NATO which was centred on 18 American-built E-3A aircraft based in West Germany. Nimrod AEW was intended to be part of a triad, the others being the fighter version of the Panavia Tornado and an improved ground-based radar chain, the so-called UKADGE. As airborne radar pickets, the Nimrods were equipped with three types of sensors--radar, target identification and ESM--six dual-screen, multi-function consoles, radio and data-link communications and, the essence of the package, the fusion of all these activities through an integrated data processing facility. Representing the visual operational capability, the consoles provide a graphics screen, alphanumeric screen, keyboard, tracker ball and special function keys for the crew of six AEW operators, each of which is appointed to his own air space jurisdiction.[3] Almost £1 billion had been expended on developing the system prior to its abandonment in December 1986, testimony enough to the sheer expense and technical complexity of electronics initiatives in contemporary defence industry and underlining--if such was needed--the risks (and potential rewards!) confronting those MIEs willing to participate in their development and production. In fact, the magnitude of the costs and the battery of technical resources required to be amassed are so daunting as to serve as significant barriers to entry into this, the most demanding, aspect of defence production. All but a few major players in the AICs are intimidated to the extent that they refrain from manufacture of defence electronics or commence it only where licensed production provides them with an entrée. China, for example, is happy to buy its more advanced avionics from the AICs: the latest fighter, the F-8-2, is to be equipped with US radar supplied by Grumman Aerospace, while the newest derivative of the MiG-21, the F-7M, is relying on equipment supplied by GEC Avionics and the standard close-support aircraft, the A-5M, is to be upgraded with an Aeritalia navigation and attack system. The country's own plants at Shanghai, Wuhan, Beijing, Nanjing and Tianjin supply less-sophisticated fire control digital computers (inspired by Soviet designs of the 1950s) besides the transistors, diodes and integrated circuits common to all defence electronics systems, but China has no option other than to import the state-of-the-art systems from the key emergent MIEs in the West.[4] NIC producers like China have made huge strides in weapons platform design and manufacturing while, at the same time, accomplishing merely hesitant steps on the road to defence electronics capability (note, though, that Israel resists such categorisation). What success they have achieved is associated with their efforts to gain missile expertise, a theme which shall be addressed at a more appropriate juncture within this chapter.

The chief purpose of the chapter, however, is to trace the emergence of a new breed of MIE, confined by and large to the AICs precisely because of its command of know-how in the field of electronics. That phenomenon has occurred against a background of changing procurement patterns in which electronics systems in their own right--radars, sonars (for undersea detection) and computers--have become desirable defence items and indispensable parts of weapons platforms. What is more, electronics has occasioned the emergence of a new weapon altogether--the missile system--as a direct result of its ability to provide means of guidance for the weapon. Unsurprisingly, then, purchases of direct electronics defence items such as avionics and indirect ones such as missiles have dented the budgets available for the procurement of hulls, airframes and other vehicles. This usurping of platform emphasis in weapons procurement has led to fundamental changes in defence industry of late. In the first place, as the British shipbuilding and ship-repair cases mentioned above testify, the state has bowed to the reality of its inadequate expertise in systems work through acceding to the acquisition of state-owned traditional MIEs by the 'emergent' private MIEs: corporations believed to be unfettered by

organisational and technological precedents and constraints. Secondly, and among the private concerns
themselves, there has been a realignment in the relative importance of prime contractors as modern MIEs
have scrambled to enter the electronics industry in
order to retain technical and market position in an
era where unbridled competition from new-entry, new-
technology firms has upset long-established suppli-
er-user relationships. Thirdly, the appearance of
emergent competitors from the ranks of electronics
firms has metamorphosed the structure of the defence
industry resulting, on the one hand, in the presence
of vigorous prime contractors, or more frequently,
subcontractors anchored to process and product innovation in electronics and, on the other, in a clutch
of conglomerates derived from a kernel of electronics expertise but which have undergone a process of
aggrandizement converting them into fully-fledged
general defence contractors. Firms such as GEC in
Britain and GE and Westinghouse Electric in America
achieved that status over an extended period during
which they transformed themselves from heavy electrical engineers into conglomerates with an electronics bent. Others of the likes of Raytheon and
Sperry came to prominence on the coat-tails of the
acceptance of defence electronics by the armed services.

Together, the firms spawned by electrical or
electronics inspiration have come to straddle the
spectrum of defence production, ironically, dominating much of the production of platforms as well as
the production of systems. In a remarkable twist,
then, the practitioners of defence production have
reverted to an industrial structure reminiscent of
that applying a century ago when the firms that saw
themselves as the repositories of metallurgical
expertise decided to branch out into comprehensive
arms-armour-warship production; which is to say, to
become makers of platforms (ships) and systems (guns
and armour) alike. Evidently, the urge among MIEs to
diversify into other aspects of defence manufacture,
partly suppressed by the technological swings transforming the traditional into the modern defence contractor, is resurging with the contemporary practice
of assembling general defence production capability
around the basis of electronics expertise. The
immediate object of this chapter is to establish the
importance of the defence electronics firms. Subsequent attention is devoted to outlining their evolution as 'emergent' MIEs and underscoring missile
manufacture as the epitome of the significance and
impact of the 'new force' in defence industry.

ELECTRONICS IN THE ASCENDANT

The indisputable tie between electronics production and defence demand is made manifest by a cursory appraisal of the effects of World War 2. Employment in the US electronics industry increased by a factor of five during the course of the war and sales exploded virtually twentyfold. Of equal significance was the munificence of government in supporting R & D and the resultant offspring of that activity, wide-ranging innovation. In the USA alone, development of radar was subsidised to the tune of $2.5 billion (in contrast to the $2 billion devoted to atomic weapons development) while $1 billion was expended on the miniaturisation of electric circuits--a necessary preliminary to microelectronics--for the express purpose of improving bomb fuzes. Other innovations of pathbreaking import stimulated by the war included loran for navigation and the digital computer for ballistic calculations.[5] The art of submarine detection, and ultimately hydrography and undersea mineral exploration, was considerably furthered by developments in sonar. In the immediate flush of victory, the lessons learned from advances in defence electronics triggered developments in pulse technology, a vital aspect of digital electronics, as well as microwave detection and enlightenment of the properties of germanium and silicon as semiconductors. For its part, research into proximity fuzes led to miniaturised transceivers, that is to say, early integrated circuits. Much of this disparate innovative energy coalesced, via the researchers of AT & T's Bell Laboratories, to produce the transistor in 1947.

Regional bias was an integral part of this upsurge in electronics activity. As reported in Chapter 2, states such as California and smaller communities, the 'defence enclaves' extending from California across the 'Sunbelt' to the Eastern Seaboard, are defence dependent by virtue of their deep commitment to aerospace work and its associated electronics industry. Indeed, it is no exaggeration to claim that the prototype of all high-tech communities, 'Silicon Valley' (Santa Clara County, California), owes its very existence to military stimulus. Moulded out of the classic symbiosis of R & D on the one hand and the transformation of the product 'conception' into the 'birth' and later stages of firm formation on the other, Santa Clara County was blessed with an institution, Stanford University, sufficiently close to, and familiar with, defence operations of both the service and

contractor kind to garner a disproportionate amount
of military research funds. One of the first beneficiaries of this linkage was Hewlett-Packard, now a
major computer manufacturer. Having beginnings in a
garage in Palo Alto in 1938 the firm was awakened by
wartime demands. Its founders--the Varian broth-
ers--innovated in 1942 the klystron tube, a crucial
ingredient in radar and microwave communications
hardware, after making use of government-funded
facilities in nearby Stanford University. On the
strength of such developments the firm rapidly grew
in size until, today, it engages in the manufacture
of computers, calculators, peripherals, test and
measurement instrumentation, medical electronic
equipment, and solid-state components of all kinds.
A sympathetic research institution combined with a
high concentration of aircraft firms together conspired to induce hectic experimentation around Palo
Alto by new enterprises eager to supply components
to the burgeoning avionics and missile markets.
Adoption of the transistor and semiconductors (initially germanium but replaced by silicon--hence the
county nickname) by the military encouraged electronics firms to locate plants in the district (for
example, Sylvania, Fairchild Camera and Instrument,
Philco-Ford, GE, Westinghouse, Itel and Kaiser) and
gave rise to additional research facilities (e.g.
IBM, ITT, Admiral, Sylvania and, in 1956, the largest of them all--Lockheed--which chose a site in the
Stanford Industrial Park). All told, a true agglomeration centre in the location-theory sense was in
the making.[6]

> The need for interaction and collaboration
> between aerospace firms and their semiconductor
> producing subcontractors grew rapidly as the
> complexity of integrated circuits increasingly
> required custom-made designs. Spatial proximity
> between the components manufacturers and the
> subsystem producers economised on R & D person-
> nel and time, and reduced communications costs.
> Since most semiconductor firms were small in
> the 1960s, they had limited ability to interact
> over long distances. Thus those new firms which
> located close to prime contractors and subsys-
> tem manufacturers had a definite advantage over
> other small firms located further away.

The California experience was replicated, with less
fanfare, elsewhere. Dotted along Route 128, the
perimeter motorway of Boston, is a similar complex
of microelectronics firms. Prominent among them are

Mitre, Raytheon and Itek, all established there as a consequence of defence-induced growth in the 1950s. Again, the key ingredient of a research institution generously endowed with defence dollars is present, although in this instance it is MIT. One of the products of the interchange was Digital Equipment Corporation, DEC, which emerged in a fashion not unlike that of Hewlett-Packard on the West Coast. The firm's founder, Ken Olsen, developed in 1957 an innovative computer (the TX-O) at the MIT Lincoln Laboratory under a military contract. This machine was subsequently 'civilianised' as DEC's first mini-computer and by 1982 the firm had capitalised on its expertise in this area to the extent of cornering 42 per cent of the market in minicomputers.[7] Smaller-scale agglomerations of electronics enterprises can be found in Texas and Florida. For the former, Fort Worth has Texas Instruments, TI, joining a group of aerospace firms which, in tandem, rejoice in the name of 'Silicon Prairie'. For the latter, the zone from Titusville to West Palm Beach encompassing Orlando, Petersburg and Tampa, houses, among others, Martin Marietta, Harris Corporation, UTC's United Space Boosters, Honeywell and Racal-Milgo in yet another cluster of aerospace and electronics MIEs.[8] In no uncertain terms, defence stimulus--while recently overshadowed by civil developments in microelectronics--continues to provide the driving force for many high-tech locations in the USA and across the AICs. Some indication of the nature and scope of that stimulus follows below.

The Significance of Defence Electronics

A few striking facts make no bones about the increasing importance of electronics in defence production. Over the ten years 1977-86 the value of defence contracts to US industry rose from $160 billion to $250 billion (in constant 1982 dollars) and, with the defence share of GNP nudging up from 5.1 per cent in 1980 to 6.7 per cent in 1986, the growth rate of defence industry as a whole outstripped that of the manufacturing sector at large. Significantly, the leading beneficiary of this stimulus was the Radio and Television Communications-Equipment Industry which depended on defence spending for 49.2 per cent of its growth in the five years following 1981.[9] A similar story can be unfolded for the UK. In the aftermath of the Falklands War, one contractor, MEL (a UK subsidiary of Philips of the Netherlands), predicted that sales of defence electronics

would skyrocket by 200 per cent in the short term
before stabilising at a growth rate of about 20 per
cent annually for the foreseeable future. The
trends appeared to confirm that prognosis: between
1980 and 1981, that is prior to the conflict, US
sales of EW equipment increased by 39 per cent while
Europe's shot up by 124 per cent. Of the total £7
billion in EW sales, avionics equipment garnered the
lion's share (65 per cent) followed by shipborne
electronics (25 per cent) and land-based systems
(ten per cent). Europe alone was expected to buy
$333 million-worth of EW equipment each year for the
succeeding half-decade.[10] According to the 1986
Defence Estimates, the expenditures on defence electronics in the UK went up, in real terms, by 36 per
cent between 1979-80 and 1983-4; a rate of growth
unmatched by any other defence industry (aerospace,
for example, increased by 15 per cent, as did ordnance; shipbuilding and repair, however, decreased
by 21 per cent and motor vehicles by ten per
cent).[11] In addition, the MoD estimated that one-
third of the output of the UK electronics industry
in general is geared to defence markets; a share
substantially smaller than the one-half applying to
the aerospace industry but significant, none the
less, when the huge actual and potential market for
consumer electronics is taken into account.[12] The
largest defence electronics industry of all, that of
the USA, has few qualms about what the future holds
for it. Considering projected defence expenditures
through to 1996, the Electronics Industries Association forecasts a slackening of demand for defence
electronics in the late-1980s but a resounding
resurgence in orders subsequently. On average,
growth in the 1990s should translate into a 1-2 per
cent annual increase in sales. As a result, the
electronics share of the defence procurement budget
is projected to rise from the 35 per cent level of
1987 to 40 per cent by 1996. In constant dollar
terms, defence will account for $48 billion-worth of
electronics sales in 1987 and $51 billion-worth in
1996: a respectable performance given the backdrop
of an overall reduction in defence procurement and R
& D from $121 billion to an anticipated $117 billion.[13]

   The fruits of all this activity are already
being harvested by the MIEs. The French state-owned
defence and electronics company, Matra, reported
that consolidated net income tripled between 1984
and 1985 (from $5 million to $16.2 million) largely
as a result of profits from its electronics activities.[14] American firms, both large and small, have

earned market laurels on the strength of their electronics sales. At the upper end of the size scale,
Westinghouse was experiencing an annual increase in
defence sales of 15-20 per cent for the first half
of the 1980s. While that rate is expected to falter
to the 5-15 per cent range for the rest of the decade, the firm still expects to earn $9 billion or so
in sales associated with AWACS between 1987 and 1989
(and conceivably much more as a corollary of
recently concluded sales of AWACS to the UK and
France). A smaller but equally vital defence electronics firm, E-Systems Incorporated, achieved a
record $926.8 million in sales in 1985 while the yet
smaller Watkins-Johnson Company scored an admirable
16 per cent return on equity with sales of $232.6
million for the same year.[15] With the prospect of
such rewards, many firms are expanding their defence
electronics offerings. This tendency affects alike
long-established electronics suppliers and firms
diversifying out of more traditional lines. In the
former category is Thorn EMI. This UK company's
electronics subsidiary has a defence portfolio
extending from radar through fuzing, computer systems, electro-optics and inertial sensors to naval
data systems. The firm supplies the world's best-
selling mortar locating radar and a variety of
radars for maritime reconnaissance aircraft; it produces proximity fuzes for missiles and computers for
analysing data extracted from submarine-detecting
multibeam sonobuoys; and, among its other offerings
in shipborne systems, it provides the data distribution facilities for all new classes of warship in
the Royal Navy. An indication of Thorn EMI's export
record is provided through its contract to supply
eleven sensor packages for the USAF B-1B bomber programme and, partly in consequence, it maintains a US
affiliate, Systron Donner Corporation of Concord,
California.

Representing the latter category is Pilkington.
This longstanding UK glass manufacturer has an electronics subsidiary, Pilkington Electro Optics, which
is charged with the manufacture of head-up displays
for fighter aircraft (an activity first undertaken
in 1967) as well as day-and-night sighting and fire
control systems for MBTs (first produced in 1972).
For Pilkington, engagement in defence electronics
complements its earlier involvement in traditional
MIE activity; which is to say, its production since
1917 of submarine periscopes and masts. The allure
of defence sales in electronics does not leave
unmoved modern MIEs either. UTC, a conglomerate with
defence interests in aircraft, aeroengines and space

vehicles, has proclaimed its intention of profiting from the SDI project. To that end, it has acquired Adaptive Optics, the use of which will enable UTC to specialise in feedback systems that allow mirrors to refocus laser beams despite refractive index variations in space. Moreover, the conglomerate has pushed its Norden Systems group into SDI fire-control work and has obtained a licence to 'militarise' DEC computers (a case of DEC products turning full-circle!).[16] In fact, UTC is merely one of many MIEs nailing its colours to the electronics mast and this is a subject to which we shall return after examining the juxtaposition of modern and emergent MIEs in founding the first great electronics-based weapon system, the guided missile.[17]

THE STIMULUS OF MISSILERY

By all accounts, rocket-propelled missiles have graced battlefields for nigh on 1,000 years but the first effective guided missiles were the German V-1 and V-2 'terror' weapons of World War 2. They ushered in the era of radio-controlled missilery, soon to be supplemented by a host of guidance mechanisms, all ultimately incumbent on advances in electronics. In turn, radio command progressed to radar command: both instances in which missiles are directed by operators positioned on the ground or in the launch platform. Radar utilisation led to active radar homing, a system which allows the missile to carry its own target-searching radar, or semi-active radar homing wherein the system permits the weapon to lock onto the energy dissipated from the radar contained within the target. For its part, infra-red homing enables the missile to zero in on a target's heat emission and is especially effective against high-performance combat aircraft. A different family of sensors was developed for inertial guidance, that is to say, the use of inertial navigators to track precise trajectories from a mobile launching platform or moving missile to a designated target. These are supplemented in cruise missiles with terrain-comparison sensors which measure the profile of the ground flown over by the missile and match it against stored information (map 'surfaces') in order to maintain true courses. If that were not enough, laser technology has been harnessed for the purpose of guiding missiles with pinpoint accuracy onto MBTs and other light reflecting targets. Guidance methods such as these, combined with nuclear warheads, led to rivalry in ballistic missile systems between the

superpowers. Beginning in the early-1950s, much
frenzied energy was expended on upgrading such strategic forces from liquid-fuelled, intermediate-range
weapons to solid-fuelled long-range multiple-warhead
vehicles (ICBMs). Less attention was directed to
tactical missile development, although in time it
came to overhaul strategic missilery as a preoccupation of the emergent MIEs. Eventually, a plethora of
guided weapons were made available for land, sea and
airborne deployment: surface-to-surface missiles,
surface-to-air missiles, anti-ship missiles and air-
to-air missiles are just a few of the terms coined
to encompass them.

From the outset, governments encouraged co-operation among suppliers in their endeavours to
produce the 'system' weapon. In the USA the univer-
sities were co-opted into providing the R & D
thrust. Thus, both the Jet Propulsion Laboratory of
the California Institute of Technology and the Aeronautical Laboratory of Cornell University were
inducted into service at an early stage. Moreover,
while the USAF preferred to rely on private MIEs for
its missile design, the other two services actively
participated in weapon development: the US Army via
its Redstone Arsenal and the US Navy via its Air
Development Center at Johnsville (a former aircraft
plant of the failed Brewster MIE) and Weapons Center
at China Lake. Progressively, however, private MIEs
were empowered to design, as well as manufacture,
new types of missiles.[18] All along in the UK, private firms had been authorised to design missiles
and they were, without exception, modern MIEs from
the aircraft industry; namely, Armstrong Whitworth,
Avro, Bristol Aeroplane, de Havilland, English Electric, Fairey, Shorts and Vickers-Armstrong. Some of
these firms created forthwith divisions specially
tasked with missile design and production, and
located them at plants removed from their main airframe sites. For example, English Electric set up a
Stevenage missile factory far from its airframe
plants around Preston, Bristol Aeroplane opened a
missile plant at Cardiff rather than use its Bris-
tol/Filton aircraft and aeroengine complex or its
Weston-super-Mare helicopter site, while Shorts
established its Guided Weapons Division at Castle-
reagh rather than in the airframe complex at Queen's
Island, Belfast.[19] Electronics suppliers acted as
subcontractors to these aircraft prime contractors,
although firms such as Ferranti, subsequently to
become leading lights as emergent MIEs, cut their
teeth during this crucial, formative period of the
1950s.

Production of missiles was entrusted to a wider-ranging set of prime contractors in the USA of the 1950s. From the rubber industry came Firestone while the automotive industry offered up Chrysler (courtesy of the Michigan Missile Plant, a GOCO factory) and, at a later date, Ford. Early representatives of the electrical/electronics sector were evident in the form of Emerson Electric and the Salt Lake City, Utah, enterprise of Sperry Univac. The fact remains, however, that the main involvement came from aircraft firms. As in Britain, modern MIEs readily established new missile divisions and plants. GD, for example, formed Convair Astronautics around a new factory constructed at Kearney Mesa near San Diego, California. The Baltimore, Maryland, firm of Martin--later to evolve into the conglomerate of Martin Marietta--abandoned aircraft construction altogether, opting instead for Thor ICBM manufacture at a new factory in Denver, Colorado, and Bullpup tactical missile production from a plant at Orlando, Florida. Similarly, Hughes Aircraft abandoned fixed-wing aircraft production for the more promising fields of missiles and electronics. All in all, aircraft firms discovered that their future viability increasingly rested on their ability to adjust to the challenges posed by the new weapon system. That they succeeded in meeting that challenge and, thereby, thwarted in the short-term the rise of the emergent MIEs thrown up by the electronics industry, speaks volumes for the resilience of the modern MIEs. This phenomenon warrants closer scrutiny.

Impact on Modern MIEs

Modern MIEs are bracketed together with emergent MIEs in contemporary missile manufacture to such a degree that the two often appear indistinguishable. Indeed, in quite a few countries the former has adopted the mantle of the latter. For example, in France the sole authority for strategic, cruise and anti-ship missile development is invested in Aérospatiale, the monolithic state concern which is descended from the 1936 nationalisation of airframe constructors. In Sweden, a combined effort on the part of Saab-Scania, a modern MIE, and Bofors, a traditional MIE, is instrumental in producing the successful RBS series of anti-ship, anti-tank and surface-to-air missiles. Israel, too, relies for much of its missile expertise on a modern MIE which has outgrown its airframe origins, that is to say,

IAI. For its part, the UK looks to two modern MIEs,
BAe (the inheritor of the numerous projects initiated by the airframe makers of the 1950s) and
Shorts, for the bulk of its missile procurement. By
virtue of the sheer size of the defence market,
though, the USA succours both kinds of MIE in profusion. Table 6.1 summarises missile prime contractors
by organisational genesis--modern or emergent--and
by product speciality. A word of caution is in order
before commenting on the table and that relates to
the organisational division of contractors. Several
of the firms are effectively general defence contractors which have expanded far beyond their origins. Thus, while placed in the emergent MIE category, LTV converted itself in reality from a minor
defence electronics firm in the 1950s to a major
electronics and airframe firm in the 1960s, reverting to an emphasis on the former in the 1980s. Currently, its function as a prime warplane contractor
has receded (recall Chapter 3) relative to its ordnance and missile interests but, perversely for ease
of classification, it continues to operate as a
major modern (both aircraft and AFVs) and emergent
MIE--and much else besides in the civil sector. Similarly, Hughes (now GM Hughes) and Martin Marietta
defy easy categorisation, and both are allotted to
the modern MIE class solely on the basis of historical circumstance: their involvement in aircraft
manufacturing having dwindled away many years ago.

Such provisos notwithstanding, from Table 6.1
can be elicited the significant fact that strategic
missiles (ICBM and SLBM types) are the preserve of
firms with aircraft-industry antecedents--Boeing,
Lockheed and Martin Marietta--whereas cruise missiles are monopolised by Boeing and GD. In short,
the 'heavy end' of the missile business is cornered
by modern MIEs. By way of contrast, the smaller SAM,
AAM and ant-tank types are readily entertained by
emergent MIEs. In the SAM type, for example, only GD
of the aircraft manufacturers maintains a product
stake but LTV, Raytheon, RCA and Ford Aerospace are
all engaged from the emergent group. Another contrast between the modern and emergent categories is
evident from the table; namely, the tendency for the
former to participate in more than one missile type
while several of the latter must make do with specialisation in only one niche. Accordingly, among the
modern MIEs, Boeing is active in four missile markets: strategic, cruise, air-to-surface and anti-
submarine; GD, GM Hughes and MD pursue three; Lockheed and Martin Marietta undertake two; and only RI
and aeroengine consortium Aerojet/Avco (the latter

part of the consortium being a component of Textron) are left with a single toehold in the business. Yet, among the emergent MIEs, no less than four--Motorola, Emerson, Gould and RCA--stick to one niche and are thus in contradistinction to LTV, TI, Raytheon and Ford Aerospace which manage to straddle three. Clearly, a scale effect is at work here. Not only do several of the emergent MIEs lack the expertise or market power to simultaneously follow several lines of development in missilery, a capability which has been acquired by a small number of their counterparts only; but not even the latter, broad-based emergent MIEs have accumulated the longstanding strength of modern MIEs of the likes of Boeing and GD.

This size advantage becomes even more apparent when the full range of suppliers involved in missile production is taken into account. Table 6.2 outlines the big-budget missile programmes under way in the USA in the mid-1980s. The three largest production ventures authorised in FY85 were the Patriot long-range air defence system for the US Army, the M-X Peacekeeper ICBM for the USAF and the Standard ship-defence SAM for the US Navy: all dominated either by modern MIEs or that quintessential broad-based emergent MIE, Raytheon.[20] While less prominent, the other programmes tabled have sizeable procurement budgets and none has less than $250 million assigned to it. Of the 15 projects in question, four have twin prime contractors (that is, a combination of 'lead' and 'co-operative' primes in the terminology introduced in Chapter 3) and two of these are combinations of emergent and modern MIEs (respectively, Raytheon with Martin Marietta and GD). A few specialist rocketry firms constitute the subcontractors delegated to providing the propulsion requirements of the missiles while the electronics industry comes into its own in providing the guidance systems. These latter fall into three kinds. The first embraces aircraft firms that are usually, but not exclusively, prime contractors for the missile airframe as well as guidance system suppliers. GD, RI, Martin Marietta and MD accomplish both roles leaving Northrop to settle for guidance work only. The second encompasses emergent firms undertaking airframe as well as guidance functions, and Raytheon and TI fit the bill here.[21] The third kind points to electronics firms content to focus on guidance systems: Honeywell, Goodyear Aerospace, IBM, Lear Siegler and Emerson Electric are examples which can be elicited from the table.[22] When the minor missile programmes are taken into consideration, the list of emergent

Table 6.1 : US Missile Prime Contractors

Missile contractor	Type ICBM SLBM	Cruise	Tactical	Air to surface	Anti ship	Anti submarine	SDI	Surface to air	Air to air	Anti tank
MODERN										
Boeing	X	X		X		X				
Lockheed	X						X			
Martin Marietta	X		X							
GD		X						X	X	
Aerojet/Avco			X							
Hughes				X					X	X
Rockwell				X						
MD					X			X		X
EMERGENT										
LTV			X				X	X		
TI				X	X					X
Motorola				X						
Emerson				X						
Gould						X				
Raytheon								X	X	X
RCA								X		
Ford								X	X	X

Source: Inferred from information in <u>Flight International</u>, 7 February 1987.

MIE subcontractors is readily extended. Litton, for instance, complements Honeywell in providing the guidance apparatus for the recently completed Boeing Air-Launched Cruise Missile programme, Singer provides a similar service for Boeing with the Short-Range Attack Missile, GE acts as one of the guidance suppliers for Lockheed's older Polaris and Poseidon SLBMs, while E-Systems and Systron Donner (Thorn EMI) provide the guidance for LTV's Lance battle-field support missile.[23] Nowhere is the symbiosis between the electronics newcomers and the modern MIEs more striking than in the massive Peacekeeper ICBM project. A veritable nexus of suppliers is required for its implementation. Martin Marietta, the representative of the modern MIEs, serves as prime contractor and reserves missile assembly and testing in its own hands. It is accompanied by other firms hailing from the aircraft industry; namely, Aerojet (second-stage booster), Avco (re-entry systems integration), Boeing (launcher), Northrop (inertial gyro) and RI (fourth-stage system). In addition, electronics firms are present: for example, GE (warhead electronics), GTE Sylvania (launch control system), Honeywell (specific force integrating receiver and second gyro), TRW (targeting and analysis program) and Westinghouse Electric (launch canister).[24]

How did the modern MIEs effect the successful transition into the missile market and achieve the mastery of electronics which such a course entailed? Ostensibly, they faced considerable barriers to entry. In the first place, they lacked the specialist technical expertise newly minted by the electronics enterprises and, secondly, their plant was ill-adapted to accommodating the needs of the process technology required for missile manufacture. A brief insight into the circumstances surrounding a missile factory of the 1980s underscores the fact that missiles require specific production facilities. The Tucson, Arizona, GOCO plant used by GM Hughes for building TOW anti-tank missiles and a variety of AAMs was shut down in 1985 owing to poor quality control--and this despite making close to one-half million missiles--and had to be thoroughly revamped before resumption of production was permitted. Costing $200 million, the rehabilitation has enabled employment to be boosted from 6,000 to 8,500.[25] It is centred on the introduction of processes of automated materials handling and computer-integrated manufacturing. Conversion of draughty, dusty, hangar-style aircraft assembly facilities into single-storey, dust-free, pristine-clean automated assembly line missile plants is either exorbitantly expensive or simply impractical. Yet, aircraft companies were prepared to risk investment both in a new technical area and in its plant appurtenances partly because other forms of diversification had proven unsatisfactory (e.g. civil aviation) and partly because the companies were aware, on the one hand, of the intricacies of dealing with the DoD and felt, on the other, that their technically-qualified staff could adjust to the new technology: feelings which were later borne out by events. As a spur to missile entry, the aircraft companies witnessed a 34 per cent decline in military aircraft sales between 1956 and 1961: in short, they had urgently to find alternative markets to staunch their eroding profits.

Consequently, their practised and well-established relations with the DoD were turned to good account in allowing them to accomplish the

Table 6.2 : Main US Missile Procurement, FY85

Type	FY85 cost ($ million)	System contractor	Propulsion contractor	Guidance contractor
Patriot	1,054	Raytheon/MM[1]	Morton Thiokol	Raytheon
Peacekeeper	991	MM	various[2]	RI/Northrop/Honeywell
Standard	754	GD	various[3]	GD
HARM	590	TI[4]	Morton Thiokol	TI
Tomahawk	581	GD/MD	Williams	MD/GD
GLCM	577	GD/MD	Williams	MD/GD
MLRS	529	Vought (LTV)	Atlantic Res	-
Phoenix	441	GM Hughes	Hercules	GM Hughes
Pershing II	382	MM	Hercules	Goodyear
IIR Maverick	378	GM Hughes	Morton/Aerojet	GM Hughes
Sparrow	363	Raytheon/GD	Hercules/Aerojet	Raytheon/GD
Harpoon	351	MD	Teledyne CAE	various[5]
Stinger	280	GD	Atlantic Res	GD
TOW-2	258	GM Hughes	Hercules/Morton	Emerson
Laser Hellfire	251	RI	Morton Thiokol	MM

Notes: 1. MM = Martin Marietta
2. Morton, Avco, Aerojet, Hercules and RI Rocketdyne
3. Aerojet, Hercules, Morton, Atlantic Research
4. Texas Instruments
5. TI, IBM, Lear Siegler and Northrop

Source: Aerospace Facts and Figures (Aerospace Industries Association of America, October 1986).

changeover from reliance on warplane orders to reliance on missile contracts as well as--or in some cases, progressively instead of--warplane manufacture. From a 1956 state of affairs in which missiles constituted a meagre 5.7 per cent of the output value of the aircraft firms, the changeover had effected a revolution by 1961 where missiles made up 44.4 per cent of the output value of those firms.[26] Table 6.3 gives some indication of the consolidation of aircraft firms in the missile market during that vital period of five years. While most of the

airframe manufacturers had some exposure to missile
work prior to 1956, the majority of them seriously
extended their missile manufacturing activities only
in the succeeding half-decade. Thus, Martin boosted
its reliance on missiles from a situation in which
they amounted to 9.7 per cent of the firm's military
sales in 1956 to one where they were responsible for
87.5 per cent of military sales in 1961. North
American Aviation (later to be absorbed into RI)
converted an insignificant missile contribution to
defence sales in 1956 into a highly respectable 40.3
per cent share in 1961. GD's Convair division more
than doubled its reliance on missile sales over the
same period. Together, these three companies
accounted for almost 40 per cent of all US missile
sales by 1961. Only Chance Vought failed to bolster
its missile stake during this period, a failing
which was to undermine the firm's independence and
led, in due course, to its acquisition by the emergent Ling-Temco defence electronics firm (creating
LTV). Fairchild, meanwhile, also courted subsequent
vulnerability by refusing to penetrate the missile
market, preferring instead to concentrate on aircraft and other activities (a situation not remedied
by the 1965 acquisition of Republic, a firm also
dangerously exposed to fickle warplane markets owing
to it being remiss in expanding its late entry into
missilery). Another late entry into missile work,
Grumman, equally learned the precariousness of over-
reliance on the vulnerable warplane market and
belatedly re-entered missile manufacture in the
1980s through its co-production of Norwegian Kongsberg Vapenfabrikk missiles. Two modern MIEs rather
tardy in entering missile production--Lockheed and
Boeing--were able, nevertheless, to become very
adept at penetrating the missile market. Their success was so precipitate, indeed, that by 1961 the
two firms between them controlled 28 per cent of the
total US missile market. All told, the aircraft
firms had captured 74.7 per cent of that market by
1961, no less than 68 per cent being in the hands of
the 'big five'; namely, Lockheed, GD's Convair division, Martin, Boeing and North American Aviation.
    The impetus gained in the late-1950s continues
to place some modern MIEs in leading missile market
positions to this day. Evidence of this is corrobo-
rated by their willingness to invest in new plant.
MD, for one, is prepared to redouble its efforts in
the defence field largely because recent attempts at
diversification into civil products have taken on
the appearance of unmitigated failures.[27] Consequently, the firm opened a Delta launch vehicle

Table 6.3 : Missilery and US Aircraft Firms

Firm	Entry year	Missile % of firm's military sales in entry year (or 1956)	Missile % of firm's military sales in 1961	Share of missile market 1961 (%)
Chance Vought	before 1956	37.7	3.9	insig.
Convair	before 1956	20.6	46.1	15.5
Douglas	before 1956	1.4	39.3	3.5
Fairchild	before 1956	insig.	0	0
Martin	before 1956	9.7	87.5	13.6
McDonnell	before 1956	2.8	18.9	1.4
North American	before 1956	insig.	40.3	10.8
Northrop	before 1956	33.7	35.5	1.2
Lockheed	1957	2.2	69.4	17.3
Boeing	1958	7.5	36.8	11.1
Beech	1960	17.5	67.5	insig.
Grumman	1961	insig.	insig.	insig.
Republic	1961	insig.	insig.	insig.

Source: Derived from G.R. Simonson, 'Missiles and Creative Destruction in the American Aircraft Industry', Business History Review, vol. 38 (1964), pp. 302-14.

assembly plant in Pueblo, Colorado, in 1987 and has plans to open a C-17 subassembly plant in Macon, Georgia, in 1989: both ventures indicative of its desire to balance investment in missile and aircraft work in the tradition of symbiosis established 20 years earlier.

However, perhaps the best example of a long-standing MIE entering and subsequently dominating the missile field is offered by Japan's MHI. This firm successively graduated from preoccupation with traditional MIE tasks (ships, ordnance and armour) through those of the modern MIE (aircraft and MBTs) to grasp the means of designing and manufacturing missiles. In short, the enterprise has progressed from the status of 'licence producer' of American missiles to becoming, at one and the same time, a licence producer and a 'lead prime'. Despite the fact that MHI builds aircraft and missiles under US licence at a cost 30 per cent higher than that prevailing in the USA, the Japanese desire for domestic production ensures the firm a guaranteed market.[28] In 1985 MHI was selected as the prime contractor responsible for building the Raytheon Patriot in Japan (with its affiliate, Mitsubishi Electric, as principal subcontractor) in a programme likely to incur costs of $3 billion. The company also makes MD's Harpoon under licence and has gone on to design its own guided weapons, most notably the AAM-1 (since 1962) and AAM-2 (since 1975) air-to-air missiles, and ASM-1 (since 1973) and SSM-1 (in the 1980s) anti-ship missiles. A conglomerate with similar antecedents, KHI, is content, for the most part, to settle for missile licence assembly, although--in the style of a 'dependent co-producer'--it is producing the Type 79 anti-tank type which is an indigenous design inspired by the GM Hughes TOW. Yet, the pre-eminence of these modern MIEs in the Japanese missile market seems assured in light of the reaction of electronics firms. NEC, for instance, through its Japan Aviation Electronics subsidiary, has long supplied guidance equipment for missiles but has never exceeded subcontractor status. Although attaining prime contractor status with the Tan-SAM, Toshiba's product proved to be such a disappointment in service that the electronics giant has been loth to take up the cudgels for missile expansion.[29] The same reluctance cannot be said to typify South Africa's Armscor, a modern MIE with considerable ambitions in the missile field. As a state enterprise, Armscor holds a brief to supply virtually all the country's defence end-products. Its activities embrace, via a clutch of subsidiaries, everything from aircraft manufacture (Atlas Aircraft), ordnance (Swartklip Products, Somchem, Naschem and Pretoria Metal Pressing), small arms and artillery (Lyttelton Engineering Works) to electro-optical systems (Eloptra) and missiles (Kentron). After initially funding French missile developments (the Cactus SAM made by Matra and Thomson-CSF), Armscor decided to engage in missile manufacture in its own right, emerging as a 'dependent co-producer'. One outcome, the V3 Kukri AAM which entered service in 1975, is an incremental improvement of Matra's Magic infra-red homing missile. Significantly, however, Armscor remains dependent on imported advanced electronics components for its EW systems: a finding suggestive of the greater technical barriers to entry in EW production compared with missile manufacture.[30]

THE DEFENCE ELECTRONICS SPRINGBOARD

Unlike in the missile market, modern MIEs have not been able to blunt the rise of emergent MIEs in the realm of EW. As hinted, such is the case because of higher technical barriers to entry. This is not to say that modern MIEs have not tried valiantly to diversify into electronics. Indeed, the acquisition of missile capability almost invariably led to involvement in electronics. It can be evinced in Figure 6.1, for example, that most US prime aircraft contractors with missile pretensions also have an accompanying electronics division. To put the stamp of approval on electronics activity, Boeing went so far as to create, in 1985, a new subsidiary, the Boeing Electronics Company, specifically geared to capturing contracts dealing with SDI spacecraft computers.[31] The airframe colossus had been involved in defence electronics engineering for many years prior to this event, albeit courtesy of its Boeing Military Airplane Company subsidiary. As a point of fact, aircraft companies began snapping up smaller electronics enterprises in earnest during the 1950s. GD, for one, bought Stromberg-Carlson, Northrop took over Page Communications while Lockheed acquired Stavid Engineering. The acquisition predilection continues: RI, for example, bought Allen-Bradley to complement its existing Collins electronics interests while no less a MIE than Lockheed felt impelled to take over Sanders Associates for $1.2 billion in 1986. As a maker of EW jammers for the USAF, Sanders offers Lockheed a defence market alternative for the late-1980s when aircraft procurement is expected to decline.[32] Other aircraft companies prefer to strengthen their defence portfolios in a similar manner. Kaman Corporation, a manufacturer of helicopters for the US Navy, took Locus under its wing in 1986 and thereby gained access to defence electronics, signal intelligence and telecommunications markets.[33] Fairchild, a company plagued with problems in its principal business of aircraft construction, has veered towards electronics as a survival ploy. Indeed, it went so far as to offer its aircraft business for sale in 1987, preferring to concentrate its energies on electronics activities. Indications of this leaning are already evident. Its American Satellite Company (Amsat) provides satellite transmissions of business information while its Space Communications Company is expected to garner government contracts.[34] At a more down-to-earth level, Fairchild's main operation in the electronics field, Fairchild Communications & Electronics

Company, won a $23 million contract in 1986 to
upgrade the fire control systems of the US Navy's
Grumman F-14 fighters. It is worth observing that
aircraft enterprises elsewhere have not been rendered immune to this tendency. Thus, in something of
a coup, BAe bought Sperry Gyroscope in 1982 from its
US parent, the Sperry Corporation, and retitled it
the BAe Bracknell Division. As such, the UK enterprise gained a Bristol facility with 200 project
engineers and designers, a Weymouth plant with an
additional 200 workers occupied in making naval control and navigation systems, and a Bracknell complex
employing 2,600 workers in the manufacture of navigation equipment, weapon control systems and telecommunications apparatus.[35] Its acquisitions tendency in this area has not abated; for, in 1987 it
purchased the West German military optics company
Steinheil-Lear Siegler for $27 million.[36] Even mod-
ern MIEs with an AFV speciality are eager to acquire
electronics expertise. United Scientific Holdings
(USH), the UK maker of Alvis armoured cars, bought
Invertron Simulators in 1987 and thereby bolstered
its role in land-system defence electronics.[37]

At the forefront of defence electronics sophistication is that branch known as EW. Its very complexity precludes easy market entry and penetration.
Combining ESM, or the monitoring of electronics
emissions of the opponent's weapons systems, with
ECM, or denying the opponent the ability to gather
and transmit signals, and ECCM, or the safeguarding
of one's own radar and radio systems in the face of
the opponent's hampering efforts, EW is centred on
three main areas of technical expertise. The first
embodies the so-called passive systems and is
expressed through the design and manufacture of
receivers: devices which are able to detect the
source, intensity and direction of signals from
radar, radio and infra-red or laser emissions. The
second focuses on active systems, that is to say,
jamming devices designed to disrupt transmissions.
The third concentrates on decoys which are usually
infra-red flares or metal foil (chaff) dispensers
used to mislead the opponent's receiver equipment.

Tables 6.4 to 6.6 attempt to pinpoint the EW
capability of the Western world's MIEs. In the first
of them, US emergent MIEs are highlighted. The
listed firms are repositories of design innovation
as well as manufacturing expertise. Several are better known as leading players in civil telecommunications or consumer electronics: as witness the presence of Honeywell, IBM, ITT, Motorola, TI and
Westinghouse. Others are more at home in defence

Figure 6.1 : US Missile, Electronics and Airframe MIEs

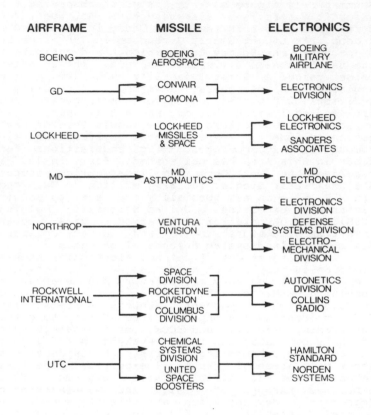

markets and E-Systems, Eaton, Loral, Raytheon and
Sperry Aerospace stand out in this respect. Regardless of overall orientation, most of the firms are
inclined to specialise within the EW nexus since
technology imposes barriers to ready diversification; indeed, only Eaton, Loral, Magnavox and TI
indulge in both receiver and jammer design. Of the
group with diverse and civil interests, ITT and
Westinghouse have formed a coalition to produce the
ALQ-165 airborne self-protection jammer for use in
A-6, AV-8, F-14, F-16 and F/A-18 aircraft.[38] The

Table 6.4 : EW and US Emergent MIEs

Firm	Capability
AEL	Receivers
Argo	Receivers
Cincinnati Electronics	Receivers
Eaton	Jammers, Receivers
Emerson Electric	Receivers
E-Systems	Receivers
ESL (TRW)	Jammers
General Instrument	Receivers
Goodyear	Decoys
Honeywell	Receivers
IBM	Receivers
ITT Avionics	Jammers
Litton	Receivers
Loral	Receivers, Jammers
Lundy Electronics	Decoys
Magnavox	Receivers, Jammers
Motorola	Jammers
Perkin-Elmer	Receivers
Raytheon	Jammers
Sperry (Honeywell)	Jammers
TI	Receivers, Jammers
Tracor	Decoys
Westinghouse Electric	Jammers

Source: M. Wilson, (ed.) Jane's Avionics 1985-86.

latter company is also renowned, of course, for its excellence in military airborne radar, particularly the AWACS type. Of the specialist defence electronics firms, Loral alone has received $100 million in EW orders from the USAF in 1986, with E-Systems and Eaton receiving, respectively, $14.5 million and $13.5 million from the same source.

Table 6.5, meanwhile, takes into its purview EW specialists located outside the USA. One of the most

Table 6.5 : EW and non-US Emergent MIEs

Firm	Country	Capability
Alkan	France	Decoys
Electronique Serge Dassault	France	Jammers
Lacroix	France	Decoys
Matra	France	Decoys
Thomson-CSF	France	Receivers, Jammers
AEG-Telefunken	West Germany	Jammers
AEL Israel Ltd	Israel	Jammers
Tadiran	Israel	Receivers, Jammers
Elettronica	Italy	Receivers, Jammers
Selenia	Italy	Receivers, Jammers
Tokyo-Keiki	Japan	Receivers
Ericsson	Sweden	Jammers
Philips Elek.AB	Sweden	Decoys
SATT	Sweden	Receivers, Jammers
GEC Marconi	UK	Receivers, Jammers
MEL	UK	Receivers
MS Instruments	UK	Receivers
Racal-Decca	UK	Receivers

Source: M. Wilson, (ed.) Jane's Avionics 1985-86.

outstanding is Thomson-CSF of France, a pioneer in laser-warning devices for helicopters. The firm is also developing an integrated ECM system for the Mirage 2000 fighter which will incorporate warning, jamming and decoying functions. Italy's Elettronica channels its attention almost solely to the EW market and employs 2,000 people near Rome making radar warning receivers and deception jammers. Since 1984 this emergent MIE has been the recipient of foreign interest with 35 per cent of its equity in the hands

of the UK's Plessey.[39] This latter, in fact,
indulges in the development of missile approach
warning systems which are a corollary of radar warning receivers. It is overshadowed in the British EW
market, however, by GEC's Marconi Defence Systems
(making receivers and jammers for RAF combat aircraft), MEL (manufacturing receivers for Army helicopters) and Racal's Radar Defence Systems group
(makers of lightweight receivers for aircraft and
helicopters).[40] Finally, Table 6.6 highlights modern
MIEs fully competitive with the emergent firms in
the EW market. Their impressive standing comes to
the fore in the competition for the avionics
required for the US warplanes of the 1990s. In so
far as EW is concerned, two teams were selected in
June 1986 to build prototype systems suitable for
installation aboard the USAF's Advanced Tactical
Fighter and the US Navy's Advanced Tactical Aircraft: the intent being to declare the winner for
the production model in 1989. One team is composed
of Lockheed's Sanders Associates and GE whereas the
other is a consortium of TRW and Westinghouse. In
other words, one of the four firms judged eligible
on the basis of technical excellence is owned by an
aircraft company; GE, conversely, is a modern MIE on
the strength of its aeroengine and marine propulsion
interests, Westinghouse likewise has longstanding
marine turbine defence interests, while TRW is a
true product of the post-World War 2 emergence of
defence electronics and systems engineering. It is
to the operations of firms in the TRW class that we
now turn.

Emergent MIE Operations

Some of the emergent MIEs juggle defence and civil
markets in attempting to ensure an element of counter-cyclical capability. A classic instance of such
motivation was the purchase of RCA by GE, effected
in 1986. The enlarged entity became a cornerstone of
the defence market--the third largest DoD contractor, in fact, with combined 1985 sales of $7.2 bil-
lion--and could count on this privileged position to
be forthcoming with 20 per cent of its aggregate
business; a proportion deemed suitable compensation
for the eroding consumer electronics market in which
GE had been battered by Japanese competition.[41] Virtually concurrently, Honeywell divested itself of
much of its loss-making civil computer operations
through the expedient of sloughing them off into a
consortium in which France's Groupe Bull and Japan's

Table 6.6 : EW and Modern MIEs

Firm	Country	Capability
IAI (Elta)	Israel	Receivers, Jammers
Mitsubishi (Electric)	Japan	Receivers, Jammers
BAe (Dynamics)	UK	Jammers
Bunker-Ramo (Allied-Bendix)	USA	Receivers
Dalmo Victor (Bell Textron)	USA	Receivers
GE	USA	Jammers
Hughes Aircraft (GM)	USA	Receivers, Jammers
Lockheed	USA	Receivers
MD	USA	Jammers
Northrop	USA	Jammers, Decoys
Sanders (Lockheed)	USA	Jammers

Source: M. Wilson, (ed.) Jane's Avionics 1985-86

NEC held the upper hand. Honeywell gambled on exchanging the recession-ridden civil computer market for the defence and aerospace market where profits are seen to be more certain. To this end, it purchased Sperry Aerospace from the newly-hatched Unisys Corporation (the merged Sperry and Burroughs) for $1.025 billion. At one fell swoop, Honeywell's annual defence electronics sales, representing about 45 per cent of its total sales, were augmented by $700 million from a 1985 level of $1.9 billion.[42] In practice, the acquisition gave Honeywell contracts for flight control systems for B-1B bomber and KC-10 tanker aircraft, visual speed indicator displays for F-15 fighters, and digital air data computers and stability augmentation systems for AV-8B V/STOL warplanes. These projects mesh well with the corporation's other defence interests which include, for example, the development of the infra-red line scanner system for the Luftwaffe's EW version of the Tornado as a combined initiative on the part of the Electro-Optics Division in Lexington, Massachusetts, and the West German subsidiary, Sondertechnik of Maintal. Equally, in the name of both synergism and counter-cyclical capability, Loral announced in 1987 its purchase for $640 million of Goodyear Aerospace, the subsidiary of the Goodyear Tire & Rubber Company. The parent firm was pressed for resources to block a takeover bid and in dispensing with its defence electronics subsidiary found the short-term relief that it was grasping for. In producing flight simulators for the F-15 and airborne parallel processors, Goodyear Aerospace was especially valued by Loral which intends to strengthen the military computer side of its offerings.[43]

A comparable desire to foment horizontal integration in defence electronics underpinned the abortive attempt made by GEC in 1985 to purchase Plessey for £1.2 billion. Unease expressed by the MoD at the prospect of a monopoly supplier (and hence an annulment of competitive tendering) in many electronics product areas was a significant factor contributing to the failure of the bid (together, in 1984-5, the firms' share of the £1.7 billion spent on UK defence electronics amounted to 73 per cent). In some contrast to their civil market fortunes, both firms had experienced impressive growth in defence sales in recent years. The Marconi Defence Systems division of GEC, for instance, evolved over the 1970s from an enterprise with sales of £12 million and a workforce of 4,000 to one notching up sales of up to £350 million and employing 13,000.[44] That same division went on to win a £550 million contract to supply heavyweight torpedoes to the MoD. The firm triumphed over Gould for the order on the promise of 5,800 new jobs by 1985 widely dispersed across the country: 500 at Hillend (Scotland), 500 at Gateshead, 250 at Leicester, 1,500 at Portsmouth, 200 at Chatham and the balance at Neston, Cheshire (the last, ironically, a redundant civil microprocessor plant). Torpedo work of the heavyweight and lightweight kind was sufficient, in fact, to keep 2,000 in employ at the Neston factory.[45] These cases illustrate the profitability presumed to derive from defence contracting, and their compelling nature in light of the alternatives cannot be gainsaid: after all, Racal sold its Bridgnorth colour TV plant to Taiwan's Tatung because it could not eradicate a £1 million annual operating loss while, simultaneously, founding an integrated circuit plant at Newbridge, near Edinburgh, to supply the needs of defence electronics and airborne navigation aids. The stark contrast between AIC vulnerability to NIC competition in consumer electronics on the one hand, and being relatively impervious to that onslaught in the field of defence electronics on the other, is wonderfully epitomised by those 1981 decisions of Racal.[46]

The Edinburgh locale for the Racal factory places it squarely in the orbit of Ferranti. This emergent MIE formed, in 1986, an Avionics Systems Group based in the Scottish capital that was aimed at the market for refitting warplanes with new avionics packages. An offshoot--Ferranti Systems Singapore--uses the NIC as a springboard for market penetration in the Far East.[47] As it is, a subsidiary in Brazil, the SFB Informatica of Rio do Janeira, manufactures computers and fire-control systems for the Brazilian Navy. In short order, Ferranti has taken under its wing the advanced laser technology branch of GE, the military trainer and air-launcher businesses of Wardle Storeys, and the laser-cutting tool company Sciaky of Chicago.[48] However, these are only the latest defence electronics ventures of the firm. Arising out of the wartime impetus given to radar and flourishing as a result of the guided weapons programmes of the 1950s and 1960s, Ferranti faltered in the 1970s and was bailed-out through part nationalisation by the UK Government (which obtained 62.5 per cent of its equity). By 1980, the government's National Enterprise Board had effectively privatised the firm, still defence-dependent to the tune of 60 per cent of its turnover. Particular assets of the firm that benefited from government cultivation were the F100L, Europe's first indigenous microprocessor, and fibre optics for use in defence optical components. Its three main divisions were the Scottish Group, Computer Systems and Electronics and, like the larger Plessey and GEC, the firm attempted to leaven the defence reliance of these units with civil work. For example, the Moston, Manchester, factory of Ferranti Instrumentation was selected in 1982 to host a joint programme with GTE (a US telecommunications giant) for the purpose of producing telephone equipment. At the same time, Moston provided the electronics distributor units for the HB876 aerial mine manufactured by Hunting Engineering for the RAF and employed 360 in this task--60 more than the workforce required for the civil telephone business.[49]

Deriving its origins as a MIE from even earlier technical developments is Sperry. We have noted that this US firm's subsidiary--a 1914 creation geared to supplying the Royal Navy with gyroscopic compasses--fell prey to the expansionism of BAe in the early-1980s. And what is more, we have made passing mention to the submerging of Sperry itself into Unisys in 1986, a restructuring which occasioned, in turn, the transfer of Sperry Aerospace to

Honeywell. At the outset, however, Sperry traded on
the excellence of its founder's gyroscope innovation
to enter defence markets in a big way during World
War 1. Fire-control apparatus, searchlights and
depth charges were all supplied to the US Navy while
the firm used the huge expansion of the aircraft
industry to justify its entry into primitive avionics: air-speed indicators, drift sets, compasses
and the like. Heightened demand in the second world
conflict confirmed Sperry's importance in gun fire-
control systems and led to its involvement in radar,
computer systems and missile guidance.[50]

In many respects the archetypal emergent MIE,
Raytheon resembled Sperry and Ferranti in its immer-
sion in defence production. Like Sperry, one stream
of its formative organisation owes its genesis to
World War 1 naval work; in this instance, primitive
sonar developments undertaken from 1917 by the Submarine Signal Company of Boston, a Raytheon division
after 1947 (Figure 6.2). In a similar fashion to
Ferranti, the US company built on its wartime
experience to develop a postwar missile expertise.
Indeed, in the 1950s, it chose to affirm its desire
to master missilery, becoming a leading prime contractor and taking over the Andover, Massachusetts,
GOCO plant from Textron and the one at Bristol, Ten-
nessee, operated by Sperry. In consequence, it prospered signally. Yet, it also gave full measure to
EW, building a dedicated ECM plant at Santa Barbara,
California, and founding a radar-making subsidiary
in Canada. Intervention in the Italian electronics
industry gave it a foothold in European missile and
defence contracting (its holding, Selenia, is now a
state undertaking under the auspices of IRI). With
the onset of the 1960s it purchased Cossor in the
UK, whereupon it gained access to further foreign
defence markets. Despite belated attempts to bolster
consumer electronics, defence still accounted for 55
per cent of Raytheon's sales in 1967 and was focused
on SLBM guidance systems, missiles and EW.[51] In
fact, diversification measures have not been overly
successful. The general aviation subsidiary, Beech
Aircraft, became a liability in the 1980s while the
energy services division and the major consumer
appliances group were performing barely adequately.
To bring home its inability to frustrate defence
reliance, Raytheon depended on EW, guidance and missiles for 60 per cent of its 1985 sales and 88 per
cent of its $376 million profits made in that year.
A single defence product, the Hawk SAM, alone had
been worth $25 billion in sales to the company in
the previous 25 years.[52]

Figure 6.2 : Raytheon's Evolution in Defence Activities

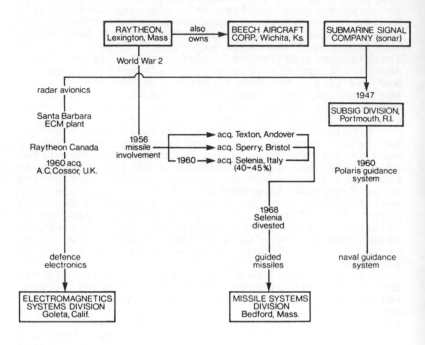

Electronics Firms Turned Conglomerate

Raytheon's acquisition of the Beech aircraft manufacturing firm as a diversification ploy hints at
the path followed by some emergent MIEs; namely, to
become multi-faceted firms that outstrip their original electronics function.[53] Firms undertake a
searching look at diversification opportunities as a
risk-reducing compulsion. When market conditions in
electronics appear less than propitious, they evince
greater interest in incursions into unrelated activities. The effect in due course is to form a conglomerate with little evident horizontal or vertical
integration. The Santa Monica, California, firm of
Lear Siegler, for example, embraced units engaged in
the manufacture of avionics (flight control systems), Smith & Wesson pistols, car seats, sail boats
and, through its Piper Aircraft Corporation

270

subsidiary, light aeroplanes. The last activity, however, proved to be such a burden to its parent, which bought it in 1984 on the promise of returns from general aviation, that it soon came to be regarded as something of a bugbear.[54] To make matters worse, the throes of takeover compelled Lear Siegler to divest itself of several electronics operations, including the sale of its Astronics and Development Sciences organisations to GEC Avionics and its Instruments and Avionics Systems Division to Smiths Industries.[55] Indeed, unrelated acquisitions entertained by electronics enterprises are frequently troublesome, and the resultant conglomerates may have difficulty sustaining their technical momentum or returning acceptable profits. The examples of LTV and Litton tend to bear out this proposition.

The former declared itself bankrupt under American Chapter 11 proceedings in July 1986. By then, it was a grouping of 65 subsidiaries around a core that had commenced as an electronics firm. In total LTV employed 56,000. Impaired over the years by ventures into meat-packing, oil-field servicing and steel-making, it was the last which pushed the firm over into bankruptcy. Aiming to dispense with the bar and tabular steelworks (though retaining the sheet-steel facilities) and the oil-service subsidiary, the conglomerate puts great faith in its defence activities as the means of restoring its fortunes. In fact, LTV Aerospace & Defense Company was the group's sole profitable operation at the time of debacle. Flush with contracts for the MLRS rocket system and the TACMS tactical missile system (to be made in a new facility at Horizon City, Texas), together with SDI contracts and airframe subcontracts for RI's B-1B, the MIE subsidiary was in no danger of faltering. Defence sales achieved an annual level of $2.5 billion with the US Government accounting for 92 per cent of the subsidiary's $4.7 billion backlog of orders at the beginning of 1986. LTV's AM General Division was also fully engaged in defence work, building military trucks. All told, aerospace and defence earned LTV a satisfying $163.9 million in 1985 on sales of $2.3 billion; a striking contrast to the conglomerate's overall operating loss of $468.7 million on sales of $8.2 billion. The hopes pinned on defence were partially realised in late-1986 when MD turned to LTV's Aircraft Products Group for the production of airframe parts for the new USAF C-17 jet transport in a deal potentially worth $3.8 billion to the conglomerate.[56]

LTV was a modest creation, the result of an

amalgamation between a minor aircraft firm and a
thrusting, but small, electronics firm. The former,
Temco Aircraft Corporation, decided to become a
prime missile contractor in the late-1950s at a time
when modern MIEs were becoming alerted to the benefits of such diversification. Accordingly, the
electronics firm of Fenske, Fedrick & Miller was
purchased by Temco in 1958 and, two years later, a
merger was effected with Ling-Altec Electronics. The
outcome was Ling-Temco Electronics, a company which
went on to acquire Chance-Vought in 1961 after that
airframe producer's attempts to enter electronics
had proved fruitless. For its part, Ling-Altec was
the descendant of Ling Electric, a firm founded by
James Ling in 1946. Ling had absorbed LM Electronics
in 1955 and Altec, a maker of sound equipment, four
years later. In the early-1960s, the MIE was reor-
ganised into three operating divisions; namely, LTV
Aerospace (with 1966 sales of $230 million), LTV
Electrosystems (1966 sales of $120 million) and LTV
Ling Altec (1966 sales of $25 million). Before the
decade was finished, however, LTV had expanded outside of the bounds of defence; albeit, its purchases
being made on the strength of profits arising out of
defence contracts. It was to amass Okonite (wire and
cables), Wilson & Company (meat-packing), National
Car Rentals, Braniff Airways and Jones & Laughlin
Steel. In the following decade the Lykes shipping
company was added while, the final straw ushering in
massive losses, was occasioned in 1984 with the
acquisition of Republic Steel. The 1986 corporate
bankruptcy--the largest heretofore in US history--
was a painful reminder to the firm of the questionable wisdom of wandering away from its MIE roots. [57]

Litton, likewise, can trace its origins to a
small electronics firm. In 1953 that firm, Electro-
Dynamics Corporation of Beverly Hills, California,
acquired a similar firm, Litton Industries, and
adopted its name. In the succeeding four years West
Coast Electronics, Ahrendt Instruments, USECO and
Automatic Sereograph all fell under the sway of Litton which, in 1958, merged with Monroe Calculator.
This merger gave Litton a civil foil to its over-re-
liance on military markets. However, while acquisitions of firms of all stripes came fast and furious
in the following several years, the firm did not
stray too far from defence involvement. Most notably, it purchased the Ingalls shipyard at Pasca-
goula, Mississippi, so as to provide warship platforms for its electronics systems and, latterly, in
1983, it bought the Lexington, Massachusetts, firm
of Itek, a leading producer of electronics equipment

used in military satellites. Its Guidance & Control
Division, in conjunction with Litton Systems Canada,
is a major supplier of inertial guidance and navigation systems for cruise missiles. Litton Industries'
sales to the US Government climbed from 32 to 57 per
cent of total sales between 1982 and 1986, with the
Ingalls Shipbuilding Division alone holding a US
Navy orders backlog worth $2.8 billion as of July
1986. On the whole, defence contributed no less than
$5.5 billion of Litton's entire backlog of $6 billion: a certain indicator of the disappointing
returns deriving from diversifications into civil
markets.[58]

Conglomerate Entry into Electronics

The boot is more often on the other foot, which is
to say, the prevailing trend is for diverse firms to
add electronics to their offerings--as witness the
examples set by modern MIEs, introduced earlier in
this chapter--rather than electronics enterprises
forcing their way into other lines of business.
Nowhere is this more evident than in the practices
of the firms that originated as heavy electrical
engineers. The US giants in the field--GE and West-
inghouse--have also, perforce, become principal
players in the defence electronics arena. For example, the former's Aerospace Electronics Systems
Department, located at Utica, New York, became a
major EW manufacturer in the early-1950s when it
supplied the USAF with 2,000 ALT-6 jammers for Boeing B-47 bombers. Thirty years later, it was still
a key factor in bomber EW; focusing on production of
components for the ALQ-161 used in the B-1B in a
programme worth in the range of $100 million annu-
ally to GE.[59] Other pioneers of electrical engineering have also responded to the siren call of defence
electronics. In West Germany, for instance, Siemens
AG is a producer of IFF recognition/identification
systems, rangefinders, passive infra-red devices and
air traffic control equipment. For its part, AEG
manufactures electronic equipment for missiles and
combat aircraft, including the Tornado, and makes
the rotor de-icing system for the US Army's MD
AH-64A helicopter. Its EW interests include jammers
and decoy transmitters, and the firm has obtained
licence rights to manufacture the GM Hughes APG-65
airborne radar for the Luftwaffe. This last project
highlights, of course, the involvement of enterprises from the automotive industry in defence electronics since both AEG--a subsidiary of

Daimler-Benz--and Hughes Aircraft--a subsidiary of GM--are owned by car makers desirous of diversifying into lucrative defence markets. Daimler-Benz had become by the mid-1980s a veritable general defence contractor, operating aircraft-, vehicle- , and marine-engine manufacturer MTU as a wholly-owned subsidiary, aircraft constructor Dornier GmbH with two-thirds of the equity, and the aforementioned AEG with a majority holding.[60]

GM, too, retains an aeroengine capability through its Allison Gas Turbine Division and long harboured airframe-making ambitions which were brought to fruition in World War 2 and the Korean War. Of late, however, it has preferred to enter the missile and defence electronics sector courtesy of Hughes Aircraft. Indeed, returns from Hughes, Electronic Data Systems and the corporation's insurance branch were more than sufficient to offset losses incurred by the mainstream car and truck businesses.[61] A fellow Detroit firm, Ford, had also dabbled in aircraft production during the phase of wartime equilibrium, but it subsequently decided to enter the peacetime defence field through the avenue of missilery. To effect its plans, Ford established an aeronautics division in 1956 and five years later absorbed TV and radio manufacturer, Philco, as a more credible agent for realising its missile ambitions. The re-styled Philco Division--later Ford Aerospace & Communications--was to gain fame as the systems contractor for the astonishingly-successful Sidewinder AAM. Unimpressed by consumer electronics prospects, the car maker sold the non-defence part of Philco to Sylvania (then a GTE affiliate) in 1974.[62] During 1986, the enterprise was to earn record contracts worth $1.7 billion; among which were awards for Sidewinder and Chaparral missiles, F/A-18 forward looking infra-red electro-optical devices, and sundry SDI projects.[63] Similar adventurous moves into defence production were followed in the UK by the Birmingham-based firm of Lucas. A specialist in supplying electrical components to the car industry, Lucas decided to expand its Lucas Aerospace division when the future of the UK automotive sector began to look grave. Consequently, it doubled the size of its aerospace interests in the late-1960s through the acquisition of units of AEI, Vickers and English Electric, concentrating throughout on the furnishing of aircraft and aeroengine control systems. An electronics department was founded in 1969 which cut its teeth on the design and manufacture of electronic engine controls for the Tornado combat aircraft. Increasingly,

electronics superseded electrical systems in the
defence products of Lucas and some labour friction
occurred as the firm attempted to rationalise its
old premises and revamp--with worker redundancies--
the reduced number required for defence production.
By the beginning of the 1980s, Lucas Aerospace was
accounting for 20 per cent of the resources of the
parent Lucas Industries, with some 50 per cent of
the subsidiary's workload stemming directly from
defence contracts.[64] A firm much smaller than Lucas
but equally committed to reliance on electronics
work for the Tornado is Frazer-Nash of Kingston-on-
Thames. To begin with, Frazer-Nash was a car manufacturer, but chameleon-like, abandoned that line to
become a dedicated MIE in World War 2 when it
designed and built gun turrets for bombers. Postwar,
the firm stayed in the defence market, developing,
for example, the weapons handling system for the
Tornado's stillborn predecessor, the TSR-2. With the
implementation of the Tornado programme, Frazer-Nash
was commissioned to work on the aircraft's gyros.[65]

A firm also guilty of revoking its ancestry is
Singer. World-renowned for both its innovation of
modern sewing machine product and process technology
and early multinational initiative (as testified
through the establishment of a model factory at Clydebank, Scotland), this venerable US company has by
and large turned its back on sewing machine manufacture. After a temporary stint of emergency involvement in defence industry in World War 2, Singer
refrained from military work until 1958 when it
bought the defence electronics firm of Haller, Ray-
mond and Brown (now, HRB-Singer of State College,
Pennsylvania). Ten years later it merged with General Precision Equipment Corporation and, in so
doing, obtained the aircraft and spacecraft Link
simulator activities. These acquisitions were
instrumental in fostering a taste for defence industry within the Singer boardroom. By the early-1980s,
some 27 per cent of its workforce of 58,000 were
employed in defence production and the firm displayed every intention of boosting that figure.
Reputedly, it was contemplating in 1986 divesting
itself altogether of its 135-year old sewing machine
and furniture activities and becoming, instead, an
emergent MIE in entirety, specialising in aerospace
and shipborne electronics.[66] As it is, defence markets account for 80 per cent of the firm's current
electronics output, the mainstay of its contemporary
operations. Singer's Kearfott Division of Little
Falls, New Jersey, is especially prominent in
defence work, producing ring laser gyro inertial

guidance systems for F-14 and F/A-18 aircraft, Trident SLBMs and M109A3 howitzers, among other weapons. Affirmation of Singer's defence orientation
came in 1986 through its purchase, first, of Dalmo
Victor from Textron for $174 million and, secondly,
of Allen Corporation of America for $20 million. The
addition of Dalmo Victor gave Singer an EW plant at
Belmont, California, whereas control of Allen provided it with access to software and training programmes complementary to its existing aircraft simulator businesses: Link Flight Simulation, Simu-Flite
and, in the UK, Singer Link-Miles.[67] In a very real
sense, then, Singer has switched from the senescence
stage in one form of industrial technology--sewing
machines--to the vibrant adolescence stage of
another; namely, defence electronics. It may be an
augury of the future for other civil-market serving
mature firms which retain sufficient pluck to act on
perceived opportunities in defence markets.

SUMMARY

In certain respects, the emergent MIE offers the
perfect vehicle for firms--provided they have technical vibrancy--that are willing to undergo the
transformation advocated by Schumpeter in terms of
'creative destruction', which is to say, they are
prepared to retrench from product lines verging on
senescence and gamble with entry into a newer technology in order to ensure survival of the corporate
entity. Many modern MIEs in the 1950s and 1960s followed that dictum precisely; becoming in the process
major missile manufacturers as well as (or instead
of) being aircraft constructors. Others, on the
fringes of modern MIE interests, such as GE and
Westinghouse Electric in the aeroengine field,
became 'lead primes' in defence electronics. The
emergent MIE is thus a hybrid: a mixture of genu-
inely new firms that innovate defence items on the
one hand and long-established defence manufacturers
diversifying into the emergent technology on the
other. Firms such as TRW characterise the former.
This Cleveland, Ohio, company sells one-third of its
output to the US Government, half of which is destined for the DoD and the other half bound for NASA.
Its speciality resides in spacecraft and missile
systems in the first place and electronics in the
second. Some 13,000 of the company's total workforce
of 85,000 are dedicated to defence contracts: in
part through the Peacekeeper ICBM contracts undertaken at Redondo Beach and San Bernardino,

California, and partly through the firm's contract
to organise the systems engineering for the Consolidated Space Center at Peterson AFB, Colorado.
Indeed, the firm owes its inception to systems engineering, managing the first USAF ICBM programme in
1954. Four years later, the firm of Ramo-Wooldridge
quit its 'childhood' stage and entered 'adolescence'
through its merger with Thompson Products. The
resultant TRW summarily downgraded USAF missile management as its primary focus, preferring direct
involvement in the manufacture of missiles, spacecraft and electronics.[68] From such small acorns,
mighty oaks do grow! MHI, meanwhile, is a salient
example of the longstanding traditional and modern
MIE which has progressed from a tentative entry into
missilery via licence production to becoming a 'lead
prime' in the production of indigenously-designed
missile systems.

It is a striking fact, none the less, that most
of the emergent MIEs dwelt upon in this chapter have
been firmly rooted in the AICs, and that is so irrespective of their origins as new entries or longstanding defence-industry participants. Technological barriers to entry suffice, in short, to
prevent the easy adaptation of NIC enterprises to
the rigorous demands of electronics-based weapons
systems. Those aspirants must settle, in the main,
for the lower end of the capability spectrum as
expressed through licence production or, as in the
case of Armscor, through incremental improvements on
licence production. Yet, Israel belies this generalisation. Its emergent MIEs have successfully accomplished the transition from licence production,
through incremental improvements on licence production and limited indigenous design, to fully-fledged
design capability in both missile and EW product
areas. Under the auspices of Rafael, the state R & D
and design authority which is sited at Haifa and
employs more than 7,000 people, the country has made
huge strides in missiles, EW, electro-optics and
thermal imaging. Among its landmark products are the
David family of artillery computers, the Shafrir 2
AAM, the Python 3 AAM and the Barak vertical-launch
anti-missile missile. IAI, the state enterprise and
modern MIE, runs Elta Electronics Industries. Strong
in active EW and military communications systems,
Elta employs 2,200 at Ashdod and achieved a 1985-6
turnover of $160 million. Another IAI subsidiary,
MBT Weapons Systems, manufactures the Gabriel anti-
ship missile, an Israeli export success story. Tan-
gentially, the woes surrounding IAI's Lavi fighter
aircraft programme might compel that modern MIE to

adopt more of the trappings of a defence-electronics concern. State enterprises do not monopolise the scene in Israel, however. Strictly emergent MIEs largely thrown up by the private sector include Tadiran, Elisra, Elbit Computers, Electro-Optics Industries, TAT, Telkoor and the Electronics Corporation of Israel: their very profusion bearing testimony to the particular environment that is conducive to the thriving of defence industry in Israel.[69] Just how much of a model this Israeli example sets for military industrialisation in other NICs is unclear, but its very uniqueness in the contemporary developing world is enough to question whether it is capable of replication.

NOTES AND REFERENCES

1. In detail, the range of production cost attributed to avionics varies from 25 to 40 per cent for European-made warplanes and 25-50 per cent for US machines. In contrast, airframe costs comprise 30-50 per cent and 25-40 per cent, respectively, of European and US aircraft production costs, with the balance ascribed to engine costs. See K. Hartley, NATO arms co-operation: a study in economics and politics, (George Allen & Unwin, London, 1983), p.103 and Appendix A. Recently, one of the successful bidders for the US Advanced Tactical Fighter development contracts, Lockheed, estimated the breakdown of that aircraft's stipulated $35 million unit costs; namely, $14.6 million for avionics (i.e. about 40 per cent of the total), $13 million for the airframe, $7.1 million for the engine and $0.3 million for armaments. See Aviation Week & Space Technology, 10 November 1986, pp.22-3.

2. R. T. Pretty (ed.), Jane's weapon systems 1983-84, (Jane's Publishing, London, 1983), pp.143-4.

3. T. G. Swan, 'Nimrod AEW: an airborne control centre' in Advances in command, control and communication systems: theory and applications, (Institution of Electrical Engineers, London, 1985), pp.220-3.

4. As mentioned in Interavia, February 1987, pp.109-10 and D. L. Shambaugh, 'China's defense industries' in P. H. B. Godwin (ed.), The Chinese defense establishment: continuity and change in the 1980s, (Westview, Boulder, 1983), pp.60-1.

5. As cogently presented by D. F. Noble in his book Forces of production: a social history of industrial automation, (Knopf, New York, 1984),

pp.7-52.
6. A. Saxenian, 'The genesis of Silicon Valley' in P. Hall and A. Markusen (eds.), Silicon landscapes, (Allen & Unwin, Winchester, Mass., 1985), pp.20-34; quote from pp.26-7. Note, the first military production venture in Santa Clara County was the conversion of FMC's agricultural tractor line at San Jose to AFV production; an activity to which it is still dedicated.
7. E. M. Rogers and J. K. Larsen, Silicon Valley fever: growth of high-technology culture, (Basic Books, New York, 1984), pp.235-8.
8. A. R. Markusen and R. Bloch, 'Defensive cities' in M. Castells (ed.), High technology, space, and society, (Sage, Beverly Hills, 1985), pp.106-20.
9. Figures calculated by Data Resources International and quoted in The Economist, 13 December 1986, p.79.
10. A. Chuter, 'Boom at the bottom end of the battle', The Engineer, (11 November 1982), pp.24-5.
11. Replicated and discussed in M. Kaldor, M. Sharp and W. Walker, 'Industrial competitiveness and Britain's defence', Lloyd's Bank Review, no. 162 (October 1986), pp.31-49.
12. See The Engineer, 14 July 1983, p.33.
13. Refer to Aviation Week & Space Technology, 20 October 1986, p.131.
14. See Flight International, 31 May 1986, p.10. Incidentally, the French Government announced its intention in 1987 to privatise Matra.
15. Note Business Week, 28 July 1986, p.31 and p.75.
16. See report in Flight International, 27 September 1986, p.34.
17. Intriguingly, UTC attempted a foray into the semiconductor field through the purchase of Mostek in 1979. In the event, the venture proved to be a perpetual loss-maker for UTC and was disposed of in 1985.
18. The US Army was forced to relinquish use of its arsenal system for missile design following pressure from the USAF and private MIEs. See C. D. Bright, The jet makers: the aerospace industry from 1945 to 1972, (Regents Press of Kansas, Lawrence, 1978), p.55. Interestingly, some countries rely on their arsenal system to both design and produce missiles, not least the Soviet bloc. An arsenal in all but name, Norway's Kongsberg Vapenfabrikk has been manufacturing ordnance since 1814 and currently produces the successful Penguin anti-ship missile--a weapon which Grumman is co-producing for the US Navy. Ironically, despite missile successes, losses

incurred in manufacturing aircraft components obliged the government to rationalise Kongsberg in 1987, a process entailing sale of the civil aeroengine division. See Flight International, 14 February 1987, p.13.

19. In the case of de Havilland, the firm's propeller-making subsidiary was charged with missile work and steadily converted its Hatfield and Lostock plants to this activity as demand for aircraft propellers dropped away in the 1950s. Refer to P. J. Birtles, Planemakers: 3. de Havilland, (Jane's Publishing, London, 1984), pp.54-8.

20. In FY87 it is envisaged that full production of the Trident II for the US Navy will be effected. Over $1.4 billion is expected to be put aside to cover these costs. See AIAA, Aerospace facts and figures 1986-1987, (Aerospace Industries Association of America, Washington, DC, October 1986), pp.48-9.

21. GM Hughes also fulfils both prime and guidance functions but it is not easily categorised in light of its historical evolution.

22. Note, however, that these firms may have attained prime contractor status in other programmes. To take but one case, Goodyear Aerospace acted as prime contractor for the Subroc anti-submarine missile, still in the US Navy inventory.

23. Although, strictly, GE is a modern MIE by dint of its supply of steam/gas turbine machinery for warships and turbine aeroengines for military aircraft.

24. See Pretty, Jane's weapon systems, pp.30-1. Also involved is specialist propulsion firm, Hercules (for the third-stage systems).

25. Note Flight International, 3 May 1986, pp.36-7.

26. G. R. Simonson, 'Missiles and creative destruction in the American aircraft industry, 1956-61', Business History Review, vol. 38 (1964), pp.302-14.

27. Its Information Systems Group recorded 1985 losses of $109.3 million on sales of $1.1 billion while its civil aircraft division showed a profit only for the fourth time since 1967. See Business Week, 28 April 1986, pp.84-6.

28. As mentioned in The Economist, 1 June 1985, p.71.

29. N. W. Davis, 'Is Japan ready for missile independence?', Aerospace America, (March 1986), pp.32-47. Note, there is a further domestic missile maker, Nissan Motors, which is also drawn from the ranks of non-electronics firms.

30. R. Leonard, South Africa at war: white power and the crisis in southern Africa, (Lawrence Hill, Westport, Conn., 1983), pp.131–59.

31. See Aviation Week & Space Technology, 23 February 1987, p.26.

32. As discussed in Business Week, 29 September 1986, p.84.

33. Note Aviation Week & Space Technology, 11 August 1986, p.73.

34. Note Business Week, 23 May 1983, pp.134–7 and 27 July 1987, p.34.

35. Sperry Gyroscope also operated a Plymouth factory employing 600 workers making Post Office equipment. Refer to The Engineer, 2 January 1982, p.11.

36. See Jane's Defence Weekly, 18 July 1987, p.110.

37. As mentioned in Jane's Defence Weekly, 11 July 1987, p.60. USH has interests in defence electronics through other ventures, including stakes in Sopelem in France and Arab International Optronics in Egypt. See International Defense Review, July 1987, p.961.

38. Detailed reports on contemporary EW projects can be found in Aviation Week & Space Technology, issues of 9 and 16 February 1987.

39. See Interavia, May 1986, p.529.

40. As listed in Interavia, April 1986, p.374.

41. See Business Week, 30 June 1986, p.64.

42. Reported in Business Week, 15 December 1986, p.13. For a review of Sperry's USAF contracts, see Aviation Week & Space Technology, 2 June 1986, p.24.

43. As noted in Aviation Week & Space Technology, 19 January 1987, p.30.

44. A number of sources are relevant here, for example, The Economist, 9 August 1986, p.51 and The Engineer, issues of 15 October 1981, p.36, and 23 January 1986, p.8.

45. See The Engineer, issues of 9 July 1981, p.10, and 10 September 1981, p.1. GEC's competitor, Gould of Chicago, plans to sell its defence activities—focusing on torpedoes and sonar—in order to buttress its civil lines of high-speed computers and industrial automation: an interesting reversal of the trend emphasised in the text. Note Business Week, 8 September 1986, p.32. Similarly disenchanted with defence electronics—this time as a fall-out of difficulties arising over the ECM system for the B-1B bomber—Eaton is to sell off its MIE functions and concentrate exclusively on civilian vehicle powertrain and controls lines. Refer to Aviation Week &

Space Technology, 2 November 1987, p.29.

46. The link can be distilled from the circumstances unfolded in The Engineer of 1/13 January 1981 and 12 February 1981.

47. As mentioned in Interavia, November 1986, p.1238.

48. Confirmed in Jane's Defence Weekly, 27 June 1987, pp.1398-9 and Flight International, 8 August 1987, p.36.

49. Elaborated upon in Flight International, 22 May 1975, p.812 and The Engineer, 11 September 1980, pp.47-9 and 29 January 1983, p.15.

50. Abstracted from T. P. Hughes, Elmer Sperry: inventor and engineer, (Johns Hopkins University Press, Baltimore, 1971).

51. O. J. Scott, The creative ordeal: the story of Raytheon, (Atheneum, New York, 1974).

52. See Business Week, 26 May 1986, pp.72-4.

53. Paradoxically, Beech is increasingly reliant on defence sales. A collapsing civil market has been shored up, in part, by a USAF order for 40 C-12F light transports worth $52 million. Rationalisation continues apace, however. The Boulder, Colorado, cryogenics department has been sold and the Selma, Alabama, plant is to be phased out, with all activities concentrated at Wichita, Kansas. See Aviation Week & Space Technology, 24 November 1986, p.71.

54. As with Beech, the Piper concern has suffered cutbacks. Its four plants have been reduced to one (at Vero Beach, Florida). See Flight International, 7 March 1987, p.14. Bowing to reality, Lear Siegler decided in 1987 to dispense with its holding in Piper altogether--a corporate revamping which also resulted in the sale of the Smith & Wesson small arms company. Interestingly, Piper was purchased by an entrepreneur with electronics antecedents: Stewart Millar. Refer to Flight International, 30 May 1987, p.3.

55. Recollect, in addition, that as well as GEC and Smiths Industries, another British company, BAe, had purchased a Lear Siegler undertaking; namely, Steinheil-Lear Siegler. For the financial background to these sales, refer to Jane's Defence Weekly, 1 August 1987, p.198 and Aviation Week & Space Technology, 10 August 1987, p.30.

56. Note Aviation Week & Space Technology issues of 11 August 1986, pp.61-4 and 27 October 1986, p.18. See also Business Week, 4 August 1986, pp.24-5.

57. For background information, refer to R. Sobel, The age of giant corporations: a

microeconomic history of American business 1914-1970, (Greenwood, Westport, Conn., 1972) and H. O. Stekler, The structure and performance of the aerospace industry, (University of California Press, Berkeley, 1965), pp.106-7. Note, in addition, The Economist, 10 January 1987, pp.68-9.

58. See K. A. Bertsch and L. S. Shaw, The nuclear weapons industry, (Investor Responsibility Research Center, Washington, DC, 1984), pp.197-204. Also Fairplay, 12 February 1987, p.10.

59. See Aviation Week & Space Technology, 7 April 1986, pp.51-2.

60. See Interavia, May 1986, pp.478-9. Note, MTU manufactures the engine for the Leopard 2 MBT and has a 40 per cent stake in Turbo-Union, makers of the RB.199 engine used in the Panavia Tornado. It also makes diesel engines for a variety of warship types, including submarines and fast attack craft.

61. See commentary in Business Week, 3 November 1986, p.36. It is worth observing that competitor Chrysler acquired, in late-1987, Electrospace Systems, a defence electronics firm located in Richardson, Texas.

62. J. B. Rae, The American automobile industry, (Twayne, Boston, 1984), pp.114-5.

63. See Jane's Defence Weekly, 20 June 1987, p.1329.

64. H. Wainwright and D. Elliott, The Lucas plan: a new trade unionism in the making?, (Allison & Busby, London, 1982).

65. See The Engineer, 19 June 1980, p.20.

66. As noted in Business Week, 3 March 1986, p.46.

67. See Aviation Week & Space Technology, 10 February 1986, p.34 and 14 April 1986, p.137.

68. Bertsch and Shaw, Nuclear weapons industry, pp.282-9 and Stekler, Structure and performance, pp.53-4.

69. See A. S. Klieman, Israel's global reach: arms sales as diplomacy, (Pergamon-Brassey's, Washington, DC, 1985) and Interavia, March 1987, pp.206-7.

Chapter Seven

CONCLUSIONS

As the foregoing account has demonstrated, defence industry is a complex assemblage of economic activities unified by a set of fairly fundamental characteristics. Above all, defence firms--wherever they are located--serve their national governments almost to the exclusion of other markets. In a word, the government is a monopsonist. It is charged, in short, not merely with setting the level of demand for defence items, but also discharges the responsibility of determining the level of technology contained within those items. The firm, for its part, has little choice but to meet the production and technology targets set by its government paymaster. This fact is fundamental to defence industry; namely, that demand management and R & D pacesetting in conjunction furnish governments with controlling rights over the operations of defence firms located in their bailiwicks. In consequence, the firms must come to terms with the peculiarities of defence markets (not overlooking the political 'pork-barrelling' which sometimes appears endemic to it) and, unsurprisingly, many of them adopt the mantle of state enterprise in order to do so. Government influence remains pervasive regardless of ownership, however.

At one extreme, the state can dictate the choice of location for the defence enterprise, either directly as in Command Economies or indirectly through legislative pressure on contractors wishing to assuage those responsible for disbursing defence budgets. For example, the Italian Government ordered 15 HH-3F helicopters from Agusta in 1987 on condition that the contractor guarantee the continuous operation of the production line at Brindisi, a city sited in the depressed Mezzogiorno region.[1] At the other extreme, state insistence on particular policies may work to undermine the prospects of

particular defence firms or, indeed, entire defence
industries. Israel's cancellation of the Lavi programme in 1987 consequent on a decision of the rul-
ing Cabinet promises not only to devastate IAI and
deny it a coveted 'lead prime' role, but will also
likely have negative impacts on an already hard-
pressed defence industry. On one count, some 7,000
workers had been dismissed by the industry in the
1985-7 period as a direct result of reductions in
the military budget--and this had occurred prior to
the Lavi cancellation.[2] Furthermore, defence firms
are increasingly looking to export options as an aid
in confronting rising costs for weapons systems.
Any restrictions on exports imposed by the state
could seriously hinder this strategy for spreading
the load of development costs and realising production economies. Sweden's attitude to defence exports
may have forced Bofors to circumvent export controls
in supplying India with artillery. The resultant
conflict between state and company can be resolved,
but the animosity created is not conducive to smooth
relations between the two parties for future domestic defence programmes. In any event, the Swedish
Government's tight arms export policy combined with
an insistence on a policy designed to maintain a
strong defence-industrial base seems to run counter
to the global reality which compels defence firms to
seek as many exports as possible.[3] Proof of the
vital importance of export orders to defence firms
is to be got from the recent happenings befalling
Dassault-Breguet. In spite of fulfilling the capacity of French 'national champion' in combat aircraft
design and manufacture, Dassault-Breguet's failure
to secure military aircraft export orders over a two
year span led, in October 1987, to the firm announc-
ing the closure of some four facilities along with
the dismissal of 1,261 employees. Earlier in the
year, the firm had dispensed with 833 engineers,
skilled workers and administrative staff.[4]

In point of fact, more stringent relations
between state and contractor are being enforced in
the USA and the UK. In a nutshell, escalation in
defence costs have compelled these governments to
respond by mandating competitive tendering and
fixed-price contracts, as well as expecting the
firms to boost their contributions to development
costs. For the British, many companies complain of
sharply reduced profit margins on defence contracts
while, for the USA, many contractors are alarmed at
the prospect of their suppliers abandoning the
defence field altogether. One percipient observer of
the US scene points to the dangers inherent in the

new policy of fixed-price contracting.[5]

> Fixed price contracts for established production programs make sense, and in some cases they can be applied effectively to development efforts. But broad application of fixed price contracts to R & D will stifle the process by discouraging the kind of technological risk-taking that has sparked past breakthroughs in military systems.

Other industry advocates have identified additional negative side-effects arising from changes in defence procurement policy. Second-sourcing, for instance, prevents a prime contractor from anticipating reasonable returns on a programme, increases instability, encroaches on proprietary expertise and, all in all, makes defence contracting a riskier business.[6] As it is, many military programmes are undertaken at inefficient levels of production. Considering average annual procurement rates for the years 1983-7, a Congressional Budget Office inquiry identified ten weapons systems that were being bought at levels insufficient to achieve a threshold denoted as the minimum economic rate; which is to say, the rate needed to sustain an acceptable rate of return for both the contractor and the government. Thus, instead of fulfilling that threshold with orders ensuring the production of 120 F-15 fighters, the procurement budget was adequate for an annual buy of only 41. Similarly, actual procurement of Tomahawk missiles fell short of the 300 required for viability by as many as 116.[7] While thirteen aircraft and missile programmes exceeded minimum economic rates of procurement (and some by a comfortable margin), the fact remains that even a superpower is faced with difficulties in supporting its defence firms.[8]

Given such a background, the approach taken in this book has been to focus on the defence firms and classify them according to functional speciality and degree of technological capability. All firms engaging in significant defence production were recast as traditional, modern or emergent MIEs.[9] Members of the first group are characterised by their predisposition to participate in production of warships and ordnance. Some of them have been involved in defence manufacture for a lengthy period and have evolved in tandem with the changing technologies obtaining for those weapons systems. A more recent batch of traditional MIEs has been thrown up the NICs, in part because technological barriers to entry remain relatively low and partly because civil technologies in shipbuilding and steelmaking can be adjusted to foster military capabilities. Modern MIEs, for their part, focus on AFV and, especially, aircraft weapons systems. In contradistinction to traditional MIEs, theirs is the condition of entry barriers: both in respect of production requirements (the difficulties of surmounting thresholds associated with production economies) and, more critically, in relation to design and development constraints. Difficulties notwithstanding, several NICs have recently attempted to overcome the hurdles standing in the way of entry and, making use of technology transfer, have persevered in their endeavours to establish modern MIEs. Yet more daunting are the barriers facing aspirants to the emergent MIE class. Centred on defence electronics, these enterprises continue to reside, by and large, in the AICs. Exceptions include a sprinkling of emergent MIEs in the USSR and Israel. However, the sophisticated high-technology industrial infrastructures conducive to the creation of emergent MIEs are likely to remain absent from most Third World countries for many years to come.

NOTES AND REFERENCES

1. As noted in Jane's Defence Weekly, 16 May 1987, p.922.
2. The reputed figure is found in Aviation Week & Space Technology, 1 June 1987, p.18. Possibly, some of these workers will be reinstated in their jobs as monies liberated from the Lavi programme are spent elsewhere.
3. Indications of the effects of this contradictory stance are noted in Jane's Defence Weekly, 10 October 1987, p.838.
4. Reported in Flight International, 10 October 1987, p.10 and Jane's Defence Weekly, 17 October 1987, p.854.
5. Donald E. Fink's editorial in Aviation Week & Space Technology, 9 March 1987, p.13.
6. See Aviation Week & Space Technology, 15 September 1986, p.89.
7. The figures are displayed in Aviation Week & Space Technology, 30 March 1987, p.77.
8. A dilemma likely to be exacerbated as America grapples with its grave budget and trade deficits in the late-1980s.
9. The definition deliberately included firms often perceived to be chiefly producers of civilian

products. In 1985-6, for instance, BAe, BS, GEC, Plessey and Rolls-Royce all received MoD contracts worth in excess of £250 million even though they are frequently viewed solely in terms of their commanding positions in civil markets.

AAM	air-to-air missile.
ACDA	Arms Control and Disarmament Agency (USA).
AEI	Associated Electrical Industries, acquired by GEC in 1967.
AEW	airborne early warning system.
AFB	Air Force Base, nomenclature used by the US Air Force.
AFNE	Astilleros Y Fabricas Navales Del Estado, the principal Argentine state shipbuilding concern with shipyards at Rio Santiago and Buenos Aires (Ensenada).
AFV	armoured fighting vehicle, the generic term for tanks and armoured cars.
AIC	advanced-industrial country.
AIDC	Aero Industry Development Center, the Taiwan aerospace enterprise which is a subsidiary of the state's Chung Shan Institute of Science and Technology.
AMRAAM	advanced medium-range air-to-air missile.
AOI	Arab Organisation for Industrialisation, the umbrella holding group for defence industry largely located in Egypt.
APC	armoured personnel carrier, a troop transport for combat zones.
ASW	anti-submarine warfare function.
AWACS	airborne warning and control system.
BAe	British Aerospace, the airframe and missile group resulting from a nationalisation initiative of 1977. Partly denationalised in 1981, the company is now wholly in the private sector.
BAT	British Aerial Transport, an early UK aircraft manufacturer.
BMARC	British Manufacture & Research Company, the UK division of Oerlikon-Bührle, which makes guns and munitions at Grantham.
BS	British Shipbuilders, the nationalised corporation created in accordance with the terms of the Aircraft & Shipbuilding Industries Act of

1977. All of its dedicated warship yards were returned to the private sector in 1985-6.

BSA   Birmingham Small Arms.

CAC   Commonwealth Aircraft Corporation Limited, a leading Australian aerospace contractor which was acquired by Hawker de Havilland in 1986 and is now trading as the latter's Victoria division.

CASA   Construcciones Aeronauticas SA, the Spanish state aerospace firm founded in 1923. It included minority shareholdings held by both Northrop of the USA and MBB of West Germany.

CITEFA   Instituto de Investigaciones Científicas y Técnicas de las Fuerzas Armadas. Argentina's Armed Forces Institute for Scientific and Technical Research.

CNIAR   Centrul National al Industriei Aeronautice Romane, the Romanian state aerospace organisation.

CNIM   Constructions Navales et Industrielles de la Méditerranée, the La Seyne shipyard facility of the French shipbuilding group Chantiers du Nord et de la Méditerranée. It is threatened with closure owing to paucity of work (end-1987).

CNR   Cantieri Navali Riuniti, Italy's state shipyard enterprise, now restyled Fincantieri-CNR.

CSSC   China State Shipbuilding Corporation.

DCN   See DTCN

DH   de Havilland, a UK aircraft constructor, now part of BAe.

disp   displacement tonnage, the yardstick for warship measurement which equates with the actual weight of water displaced by the vessel.

DoD   Department of Defense (USA).

DTCN   La Direction Technique des Constructions Navales, the French state dockyard organisation. Now restyled DCN or Direction des Constructions Navales.

dwt   deadweight tonnage, the means of measuring tankers and bulk cargo ships; it monitors the actual cargo carrying capacity of the ship.

ECCM   electronic counter-countermeasures.

ECM	electronic countermeasures.
EFIM	Ente Partecipazioni e Finanziamento Industria Manifatturiera, an Italian state holding company which embraces the Breda ordnance company and the Agusta aircraft firm as well as OTO Melara.
EMBRAER	Empresa Brasileira de Aeronáutica SA, the Brazilian aerospace company.
ENAER	Empresa Nacional de Aeronautica, Chile's aircraft enterprise.
ENGESA	Engesa Engenheiros Especializados SA, the principal Brazilian military vehicles manufacturer. In 1987 it took a major stake in Helibras Helicopteros do Brasil SA.
ESM	electronic support measures.
EW	electronic warfare.
F + W	Eidgenössisches Flugzeugwerk, the Swiss Government's Federal Aircraft Factory at Emmen.
FMC	Food Machinery and Chemical Corporation, a major US manufacturer of AFVs and ordnance.
FN	Fabrique Nationale Herstal SA, the leading Belgian arms manufacturer and engineering concern.
GAF	Government Aircraft Factories, the Australian state aircraft enterprise with facilities at Avalon (F-18 assembly) and Fisherman's Bend. Retitled Aerospace Technologies of Australia Pty Ltd on 1 July 1987.
GD	General Dynamics Corporation.
GDP	Gross Domestic Product, the value of goods and services produced in an economy exclusive of returns from assets invested abroad.
GE	General Electric Company (USA).
GEC	General Electric Company (UK).
GIAT	Groupement Industriel des Armements Terrestres, the French state undertaking responsible for land armaments.
GKN	Guest, Keen and Nettlefords, a UK engineering concern.
GNP	Gross National Product, an extension of GDP to include the value of all goods and services earned by a country's nationals whether at home or abroad.
GOCO	government-owned contractor-operated,

a plant funded by the US Government and run by private defence contractors.

GTE General Telephone and Electronics Corporation, a US telecommunications operator and electronics equipment manufacturer based at Stamford, Connecticut. Sylvania is a subsidiary.

HAL Hindustan Aeronautics Limited, the Indian state aerospace enterprise founded in 1964 with the merger of two smaller state aircraft undertakings.

HDW Howaldtswerke-Deutsche Werft AG, shipbuilders and repairers at Kiel and Hamburg.

IAI Israel Aircraft Industries Ltd, the main state-owned defence contractor in Israel. Formerly known as Bedek, it assumed its current title in 1967.

ICBM inter-continental ballistic missile.

IHI Ishikawajima-Harima Heavy Industries Co Ltd.

IPTN Nusantara Aircraft Industry, formerly PT Nurtanio, the Indonesian state aeronautical undertaking.

IRI Istituto per la Ricostruzione Industriale, a large Italian state holding group which includes such MIEs as Aeritalia (aircraft), Fincantieri (warships), Selenia and Elsag (both electronics).

KAL Korean Air Lines, the flag-carrier of South Korea that also undertakes aircraft assembly.

KHI Kawasaki Heavy Industries Ltd.

LTV Ling-Temco-Vought, a US aerospace company based in Dallas, Texas.

MBB Messerschmitt-Bölkow-Blohm GmbH, the leading aerospace (and courtesy of its Krauss-Maffei subsidiary, MBT and AFV) enterprise in West Germany.

MBT main battle tank, the heaviest and best armed class of tank.

MD McDonnell Douglas Corporation. The MIE is also known as McAir.

MHI Mitsubishi Heavy Industries Ltd.

MIC Military-Industrial Complex, the association between defence industry, government and the armed services, especially in the USA.

MIE military-industrial enterprise, a

	private or state firm, or division of a firm, that is primarily engaged in defence production.
MKEK	Makina ve Kimya Endustisi Kurumu, a Turkish aircraft constructor of the 1950s.
MLRS	multiple-launch rocket system.
MoD	Ministry of Defence (UK).
MTU	Motoren und Turbinen-Union München, the aeroengine subsidiary of Daimler-Benz.
NASA	National Aeronautics and Space Administration.
NATO	North Atlantic Treaty Organisation.
nec	not elsewhere classified.
NIC	newly-industrialising country.
OTO	Cantieri Odero Terni Orlando, an Italian defence contractor (now known as OTO Melara) originating as a ship-building concern.
P & WC	Pratt & Whitney Canada Inc, the Canadian aeroengine subsidiary of Pratt & Whitney; itself a subsidiary of United Technologies Corporation.
R & D	Research and Development.
RCA	Radio Corporation of America, now part of GE.
RI	Rockwell International Corporation.
RO	Royal Ordnance, the successor to the UK state Royal Ordnance Factories which, earmarked for privatisation, was bought by BAe in 1987.
ROF	Royal Ordnance Factory(ies).
SABCA	Société Anonyme Belge de Constructions Aéronautiques, the principal aerospace enterprise in Belgium. Founded in 1920.
SAFAT	Societe Anonyme Fiat Armamente Torino.
SAI	Singapore Aircraft Industries, a government-owned aerospace company founded in 1982.
SAM	surface-to-air missile.
SDI	Strategic Defense Initiative, the US 'Star Wars' space defence project.
SECN	La Sociedad Espanola de Construccion Naval.
SEPECAT	Société Européenne de Production de l'Avion ECAT, an Anglo-French organisation formed in 1966 to create the Jaguar strike aircraft.
SIA	Societa Italiano Aviazione.

SIMA	Empresa Publica Servicios Industriales de la Marina, the ship construction and repair enterprise of Peru.
SIPRI	Stockholm International Peace Research Institute.
SIT	Società Italiana Transaerea.
SLBM	submarine-launched ballistic missile.
SNC	A Quebec engineering group, owners of Canadian Arsenals Ltd.
SNECMA	Société Nationale d'Étude et de Construction de Moteurs d'Aviation, the French state aeroengine manufacturer.
SONACA	Société Nationale de Construction Aérospatiale, the joint state-private Belgian aerospace concern formed out of the ashes of Fairey SA.
SVA	Aviation Department of Ansaldo, the erstwhile Italian shipbuilder.
TI	Texas Instruments Inc.
TOW	tube-launched, optically-tracked, wire-guided missile.
TRW	Thompson-Ramo-Wooldridge, an electronics and aerospace firm based in Cleveland, Ohio.
TUSAS	Tusas Türk Ucak San AS, or Turkish Aircraft Industries.
UKADGE	United Kingdom Air Defence Ground Environment radar system.
USH	United Scientific Holdings plc, a UK group with interests in AFV production (courtesy of Alvis of Coventry) and defence electronics. Thorn EMI holds a small stake in the firm.
UTC	United Technologies Corporation.
VSEL	Vickers Shipbuilding and Engineering Ltd, the nationalised (1977) shipbuilding and ordnance assets of Vickers at Barrow-in-Furness which were returned to the private sector in 1986.
V/STOL	vertical or short take-off and landing. The flight characteristics of such specialised combat aircraft as the BAe Harrier.

References

Abernethy, W. J., Clark, K. B. and Kantrow, A. M. (1983) Industrial renaissance: producing a competitive future for America, Basic Books, New York

Abolfathi, F. (1980) 'Threats, public opinion, and military spending in the United States, 1930-1990' in P. McGowan and C. W. Kegley (eds.), Threats, weapons, and foreign policy, Sage, Beverly Hills, pp.83-133

Adams, G. (1982) The politics of defense contracting: the iron triangle, Transactions Books, New Brunswick

Adams, R. J. Q. (1978) Arms and the wizard, Texas A & M University Press, College Station

AIAA, (1986) Aerospace facts and figures 1986-1987, Aerospace Industries Association of America, Washington, DC

Alexander, K. J. W. and Jenkins, C. L. (1970) Fairfields: a study of industrial change, Allen Lane, London

Allen, G. C. (1929) The industrial development of Birmingham and the Black Country 1860-1927, George Allen & Unwin, London

Ames, E. and Rosenberg, N. (1970) 'The Enfield Arsenal in theory and history' in S. B. Saul (ed.), Technological change: The United States and Britain in the Nineteenth Century, Methuen, London, pp.99-119

Anderton, C. H. and Isard, W. (1985) 'The geography of arms manufacture' in D. Pepper and A. Jenkins (eds.), The geography of peace and war, Basil Blackwell, Oxford, pp.90-104

Art, R. J. (1968) The TFX decision: McNamara and the military, Little, Brown & Co, Boston

Auer, J. E. (1973) The postwar rearmament of Japanese maritime forces, 1945-71, Praeger, New York

Augustine, N. R. (1975) 'One plane, one tank, one ship: trend for the future?', Defense Management Journal, April, pp.34-40

Ayres, R. U. (1984) The next industrial revolution: reviving industry through innovation, Ballinger, Cambridge, Mass

Bainbridge, T. H. (1939) 'Barrow in Furness: a population study', Economic Geography, vol. 15, pp.379-83

Bajusz, W. (1981) 'International arms procurement,

multiple actions, multiple objectives' in M. Edmonds (ed.), International arms procurement: new directions, Pergamon, New York, pp.188-216

Barnaby, F. and ter Borg, M. (eds.) (1986) Emerging technologies and military doctrine, Macmillan, London

Barnes, C. H. (1964) Bristol aircraft since 1910, Putnam, London

Barnes, F. (1951) Barrow and district, James Milner, Barrow

Barry, P. (1863) Dockyard economy and naval power, Sampson Low, London

Bertsch, K. A. and Shaw, L. S. (1984) The nuclear weapons industry, Investor Responsibility Research Center, Washington, DC

Bezdek, R. H. (1975) 'The 1980 economic impact--regional and occupational--of compensated shifts in defense spending', Journal of Regional Science, vol. 15, pp.183-97

Birtles, P. J. (1984) Planemakers: 3. de Havilland, Jane's Publishing, London

Blake, J. W. (1956) Northern Ireland in the Second World War, HMSO, Belfast

Bluestone, B., Jordan, P. and Sullivan, M. (1981) Aircraft industry dynamics: an analysis of competition, capital, and labor, Auburn House, Boston, Mass

Boddy, M. and Lovering, J. (1986) 'High technology industry in the Bristol sub-region: the aerospace/defence nexus', Regional Studies, vol. 20, pp.217-31

Bolton, R. E. (1966) Defense purchases and regional growth, The Brookings Institution, Washington, DC

Brassey, T. A. (1892) The naval annual, Brassey's Publishers, London

Bright, C. D. (1978) The jet makers: the aerospace industry from 1945 to 1972, Regents Press of Kansas, Lawrence

Brown, D. K. (1983) A century of naval construction: the history of the Royal Corps of Naval Constructors, Conway Maritime Press, London

Brzoska, M. (1983) 'The Federal Republic of Germany' in N. Ball and M. Leitenberg (eds.), The structure of the defense industry, St Martin's Press, New York, pp.111-39

_____ and Ohlson, T. (eds.) (1986) Arms production in the Third World, Taylor & Francis, London

Buckberg, A. (1965) 'Federal government expenditures: a federal viewpoint', The Western Economic Journal, vol. 3, pp.126-33

References

Burns, R. D. (1972) The international trade in armaments prior to World War II, Garland, New York

Chillon, J., Dubois, J-P. and Wegg, J. (1980) French postwar transport aircraft, Air-Britain, Tonbridge

Chuter, A. (1982) 'Boom at the business end of the battle', The Engineer, 11 November, pp.24-5

Clarke, J. F. (1977) Power on land and sea: 160 years of industrial enterprise on Tyneside, Smith Print Group, Newcastle

Colledge, J. J. (1969-70) Ships of the Royal Navy 2 vols., David & Charles, Newton Abbot

Connon, P. (1984) An aeronautical history of the Cumbria, Dumfries and Galloway region, Part 2: 1915 to 1930, St Patrick's Press, Penrith

Cooling, B. F. (1979) Gray steel and blue water navy, Archon Books, Hamden, Conn

Crowell, J. F. (1920) Government war contracts, Oxford University Press, New York

Cypher, J. (1981) 'The basic economics of rearming America', Monthly Review, vol. 23, June, pp.11-27

Dahlman, C. J. (1984) 'Foreign technology and indigenous technological capability in Brazil' in M. Fransman and K. King (eds.), Technological capability in the Third World, Macmillan, London, pp.317-34

Davis, N. W. (1986) 'Is Japan ready for missile independence?', Aerospace America, March, pp.32-47

DeGrasse, R. W. (1983) 'Military spending and jobs', Challenge, vol. 26, July-August, pp.4-15

—————— (1983) Military expansion economic decline: the impact of military spending on US economic performance, M. E. Sharpe, Armonk, NY

Dethomas, B. (1984) 'Creusot-Loire—end of a 200-year industrial saga', Manchester Guardian Weekly, 15 July, p.12

Dillon, C. H. (1966) 'Government purchases and depressed areas' in H. A. Cameron and W. Henderson (eds.), Public finance—selected readings, Random House, New York, pp.97-105

Dodds, M. H. (1930) A history of Northumberland, vol. XIII, Andrew Reid, Newcastle

Donne, M. (1981) Leader of the skies: Rolls-Royce, Frederick Muller, London

Dörfer, I. (1983) Arms deal: the selling of the F-16, Praeger, New York

Dougan, D. (1968) The history of North East shipbuilding, George Allen & Unwin, London

Dyckman, J. W. (1964) 'Some regional development issues in defense program shifts', Journal of

Peace Research, vol. 1, pp.191-203

Edmonds, M. (ed.) (1981) _International arms procurement: new directions_, Pergamon, New York

Evans, H. (1978) _Vickers: against the odds 1956-1977_, Hodder & Stoughton, London

Foss, C. F. (ed.) (1985) _Jane's armour and artillery 1985-86_, Jane's Publishing, London

Gansler, J. (1980) _The defense industry_, MIT Press, Cambridge, Mass

Gold, D. (1984) 'Conversion and industrial policy' in S. Gordon and D. McFadden (eds.), _Economic conversion: revitalizing America's economy_, Ballinger, Cambridge, Mass., pp.191-203

Goodwin, J. (1985) _Brotherhood of arms: General Dynamics and the business of defending America_, Times Books, New York

Gordon, J. R. (1983) 'NATO industrial preparedness' in L. D. Olvey, H. A. Leonard and B. E. Arlinghaus (eds.), _Industrial capacity and defense planning_, D. C. Heath, Lexington, Mass., pp.35-63

Gorgol, J. F. (1972) _The military-industrial firm: a practical theory and model_, Praeger, New York

Graham, T. W. (1984) 'India' in J. E. Katz (ed.), _Arms production in developing countries_, D. C. Heath, Lexington, Mass., pp.157-91

Grant, A. (1950) _Steel and ships: the history of John Brown's_, Michael Joseph, London

Groth, C. (1981) 'The economics of weapons coproduction' in M. Edmonds (ed.), _International arms procurement: new directions_, Pergamon, New York, pp.71-83

Gunston, B. (1980) _The plane makers_, Basinghall Books, London

────────── (1986) _World encyclopaedia of aero engines_, Patrick Stephens, Wellingborough

Hahn, B. (1986) 'Chinese navy: first destroyer construction programme 1968-85', _Navy International_, vol. 91, November, pp.690-5

Hammond, P. Y. et al (1983) _The reluctant supplier: US decision-making for arms sales_, Oelgeschlager, Gunn & Hain, Cambridge, Mass

Hannah, L. (1976) _The rise of the corporate economy_, Methuen, London

Harrigan, K. R. (1983) _Strategies for vertical integration_, D. C. Heath, Lexington, Mass

Hartley, K. (1981) 'The political economy of NATO defense procurement policies' in M. Edmonds (ed.), _International arms procurement: new directions_, Pergamon, New York, pp.98-114

────────── (1983) _NATO arms co-operation: a study in economics and politics_, George Allen &

Unwin, London
_____ (1985) 'UK defence policy: seeking better value for money' in RUSI and Brassey's Defence Yearbook 1985, Brassey's Defence Publishers, London, pp.105-119
_____ (1987) 'Public procurement and competitiveness: a community market for military hardware and technology', Journal of Common Market Studies, vol. 25, pp.237-47
_____ and Corcoran, W. J. (1975) 'Short-run employment functions and defence contracts in the UK aircraft industry', Applied Economics, vol. 7, pp.223-33
Hayward, K. (1983) Government and British civil aerospace: a case study in post-war technology policy, Manchester University Press, Manchester
Head, R. G. (1974) 'The weapons acquisition process: alternative national strategies' in F. B. Horton, A. C. Rogerson and E. L. Warner (eds.), Comparative defense policy, Johns Hopkins University Press, Baltimore, pp.412-25
Herrick, W. R. (1966) The American naval revolution, Louisiana State University Press, Baton Rouge
Higham, R. (1968) 'Quantity vs. quality: the impact of changing demand on the British aircraft industry', Business History Review, vol. 42, pp.443-68
_____ (1972) Air power: a concise history, St Martin's Press, New York
Hoagland, J. H. (1978) 'The US and European aerospace industries and military exports to the less developed countries' in U. Ra'anan, R. L. Pfaltzgraff and G. Kemp (eds.), Arms transfers to the Third World: the military buildup in less developed countries, Westview, Boulder, pp.213-27
Holloway, D. (1983) The Soviet Union and the arms race, Yale University Press, New Haven
Hornby, W. (1958) Factories and plant, HMSO, London
Hounshell, D. A. (1985) 'Ford Eagle Boats and mass production during World War I' in M. R. Smith (ed.), Military enterprise and technological change: perspectives on the American experience, MIT Press, Cambridge, Mass., pp.175-202
Hoyt, E. P. (1971) The space dealers, John Day Company, New York
Hughes, T. P. (1971) Elmer Sperry: inventor and engineer, Johns Hopkins University Press, Baltimore
Huisken, R. (1983) 'Armaments and development' in H. Tuomi and R. Väyrynen (eds.), Militarization and arms production, Croom Helm, London,

pp.3-25

Humble, R. (1986) _The rise and fall of the British navy_, Queen Ann Press, London

Hume, J. R. and Moss, M. S. (1979) _Beardmore--the history of a Scottish industrial giant_, Heinemann, London

Irving, R. J. (1975) 'New industries for old? some investment decisions of Sir W. G. Armstrong, Whitworth & Co Ltd, 1900-1914', _Business History_, vol. 17, pp.151-68

Johnson, R. W. (1895) _The making of the Tyne_, Walter Scott, London

Johnston, R. J. (1978) 'Political influences on the allocation of federal money to local environments', _Environment and Planning A_, vol. 10, pp.691-704

_____ (1979) 'Congressional committees and the inter-state distribution of military spending', _Geoforum_, vol. 10, pp.151-62

Jones, R. W. and Hildreth, S. A. (1984) _Modern weapons and Third World powers_, Westview, Boulder

Kaldor, M. (1980) 'Technical change in the defence industry' in K. Pavitt (ed.), _Technical innovation and British economic performance_, Macmillan, London, pp.100-121

_____ (1981) _The baroque arsenal_, Hill & Wang, New York

_____, Sharp, M. and Walker, W. (1986) 'Industrial competitiveness and Britain's defence', _Lloyd's Bank Review_, no. 162, October, pp.31-49

Katz, J. E. (ed.) (1986) _The implications of Third World military industrialization_, Lexington Books, Lexington, Mass

Kennedy, G. (1983) _Defense economics_, Duckworth, London

Klein, B. H. (1977) _Dynamic economics_, Harvard University Press, Cambridge, Mass

Klieman, A. S. (1985) _Israel's global reach: arms sales as diplomacy_, Pergamon-Brassey's, Washington, DC

Knaack, M. S. (1978) _Encyclopedia of US Air Force aircraft and missile systems_, Office of Air Force History, Washington, DC, vol. 1

Kobayashi, U. (1922) _Military industries of Japan_, Oxford University Press, New York

Kolodziej, E. A. (1980) 'Determinants of French arms sales: security implications' in P. McGowen and C. W. Kegley (eds.), _Threats, weapons, and foreign policy_, Sage, Beverly Hills, pp.137-75

Kuhn, R. L. (1982) _Mid-sized firms: success strategies and methodology_, Praeger, New York

Kurth, J. R. (1973) 'Why we buy the weapons we do',

References

Foreign Policy, no. 11, Summer, pp.33-57

Lancaster, J. Y. and Wattleworth, D. R. (1977) *The iron and steel industry of West Cumberland: a historical survey*, British Steel Corporation, Workington

Law, C. M. (1983) 'The defence sector in British regional development', *Geoforum*, vol. 14, pp.169-84

Leonard, R. (1983) *South Africa at war: white power and the crisis in southern Africa*, Lawrence Hill, Westport, Conn

Lischka, J. R. (1977) 'Armor plate: nickel and steel, monopoly and profit' in B. F. Cooling (ed.), *War, business, and American society: perspectives on the military-industrial complex*, Kennikat, Port Washington, NY, pp.43-58

Lock, P. (1986) 'Brazil: arms for export', in M. Brzoska and T. Ohlson (eds.), *Arms production in the Third World*, Taylor & Francis, London, pp.79-104

Lonie, A. A. and Begg, H. J. (1979) 'Comment: further evidence of the quest for an effective regional policy 1934-1937', *Regional Studies*, vol. 13, pp.495-500

Loose-Weintraub, E. (1984) 'Spain's new defence policy: arms production and exports' in SIPRI, *World armaments and disarmament*, Taylor & Francis, London, pp.137-49

Lynch, T. G. (1986) 'DELEX: The Nipigon experience', *Navy International*, vol. 91, May, pp.301-303

Lyon, H. (1977) 'The relations between the Admiralty and private industry in the development of warships' in B. Ranft (ed.), *Technical change and British naval policy 1860-1939*, Hodder & Stoughton, London, pp.37-64

MacDougall, P. (1982) *Royal Dockyards*, David & Charles, Newton Abbot

Malecki, E. J. (1984) 'Military spending and the US defense industry: regional patterns of military contracts and subcontracts', *Environment and Planning C*, vol. 2, pp.31-44

Markusen, A. (1985) 'The military remapping of the United States', *Built Environment*, vol. 11, pp.171-80

_____ and Bloch, R. (1985) 'Defensive cities: military spending, high technology, and human settlements' in M. Castells (ed.), *High technology, space, and society*, Sage, Beverly Hills, pp.106-20

Massey, D. and Meegan, R. (1982) *The anatomy of job loss*, Methuen, London

Melman, S. (1985) *The permanent war economy:*

American capitalism in decline, 2nd edn, Simon & Schuster, New York

Miernyck, W. H. (1965) Elements of input-output economics, Random House, New York

Miller, J. P. (1949) Pricing of military procurements, Yale University Press, New Haven

Molson, K. M. and Taylor, H. A. (1982) Canadian aircraft since 1909, Putnam, London

Moore, J. (ed.) (1986) Jane's fighting ships 1986-87, Jane's Publishing, London

Moseley, H. G. (1985) The arms race: economic and social consequences, Lexington Books, Lexington, Mass

Naumann, W. L. (1977) The story of Caterpillar Tractor Company, Newcomen Society, New York

Nemecek, V. (1986) The history of Soviet aircraft from 1918, Willow Books, London

Neuman, S. G. (1984) 'Third World arms production and the global arms transfer system' in J. E. Katz (ed.), Arms production in developing countries, D. C. Heath, Lexington, Mass., pp.15-37

——————— (1985) 'Countertrade, barter, and countertrade: offsets in the international arms market', Orbis, vol. 29, pp.183-213

Noble, D. F. (1984) Forces of production: a social history of industrial automation, Knopf, New York

Nolan, J. E. (1986) Military industry in Taiwan and South Korea, Macmillan, London

ÓhUallacháin, B. (1984) 'Input-output linkages and foreign direct investment in Ireland', International Regional Science Review, vol. 9, pp.185-200

Pearton, M. (1982) The knowledgeable state: diplomacy, war and technology since 1830, Burnett Books, London

Perry, W. and Weiss, J. C. (1986) 'Brazil' in J. E. Katz (ed.), The implications of Third World military industrialization, Lexington Books, Lexington, Mass., pp.103-117

Pierre, A. J. (1982) The global politics of arms sales, Princeton University Press, Princeton

Pollard, S. (1951) 'The economic history of British shipbuilding 1870-1914', unpublished PhD thesis, University of London

Pool, B. (1966) Navy Board contracts 1660-1832, Longman, London

Postan, M. M. (1952) British war production, HMSO, London

Pretty, R. T. (ed.) (1983) Jane's weapon systems 1983-84, Jane's Publishing, London

Pugh, P. (1986) The cost of seapower: the influence

of money on naval affairs from 1815 to the present day, Conway Maritime Press, London

Rae, J. B. (1984) The American automobile industry, Twayne, Boston

Ram, V. S., Sharma, N. and Nair, K. K. P. (1976) Performance of public sector undertakings, Economic and Scientific Research Foundation, New Delhi

Ramamurti, R. (1985) 'High technology exports by state enterprises in LDCs: the Brazilian aircraft industry', The Developing Economies, vol. 33, pp.254-80

Rasmussen, P. N. (1956) Studies in inter-sectoral relations, Einar Harcks Forlag, Copenhagen

Rich, M. et al (1981) Multinational coproduction of military aerospace systems, Rand, Santa Monica, Calif

Riefler, R. F. and Downing, P. B. (1968) 'Regional effect of defense effort on employment', Monthly Labor Review, July, pp.1-8

Robertson, B. (1979) British military aircraft serials, Patrick Stephens, Cambridge

Rogers, E. M. and Larsen, J. K. (1984) Silicon Valley fever: growth of high-technology culture, Basic Books, New York

Rosenberg, N. (1982) Inside the black box: technology and economics, Cambridge University Press, Cambridge

Ross, N. S. (1952) 'Employment in shipbuilding and ship-repairing in Great Britain', Journal of the Royal Statistical Society, Series A, vol. 115, pp.524-33

Rubin, S. M. (1986) The business manager's guide to barter, offset and countertrade, EIU Special Report no. 243, Economist Intelligence Unit, London

Salter, M. S. and Weinhold, W. A. (1979) Diversification through acquisition: strategies for creating economic value, Free Press, New York

Saul, S. B. (1970) 'The market and the development of the mechanical engineering industries in Britain, 1860-1914' in S. B. Saul (ed.), Technological change: The United States and Britain in the nineteenth century, Methuen, London, pp.141-70

Saville, A. W. (1977) 'The naval military-industrial complex, 1918-41' in B. F. Cooling (ed.), War, business, and American society: historical perspectives on the military-industrial complex, Kennikat, Port Washington, NY, pp.105-117

Sawyer, L. A. and Mitchell, W. H. (1985) The Liberty Ships, 2nd edn, Lloyd's of London Press,

Colchester

Saxenian, A. (1985) 'The genesis of Silicon Valley' in P. Hall and A. Markusen (eds.), Silicon landscapes, Allen & Unwin, Winchester, Mass, pp.20-34

Schlaifer, R. (1950) Development of aircraft engines, Graduate School of Business Administration, Harvard University, Boston

Scott, H. F. and Scott, W. F. (1981) The armed forces of the USSR, Westview, Boulder

Scott, J. D. (1962) Vickers: a history, Weidenfeld & Nicolson, London

Scott, O. J. (1974) The creative ordeal: the story of Raytheon, Atheneum, New York

Seagram, M. (1986) 'Does relatively high defence spending necessarily degenerate an economy?', Journal of the RUSI for Defence Studies, vol. 131, pp.45-9

Sekigawa, E. (1974) Pictorial history of Japanese military aviation, Ian Allan, London

Shambaugh, D. L. (1983) 'China's defense industries: indigenous and foreign procurement' in P. H. B. Gordon (ed.), The Chinese defense establishment: continuity and change in the 1980s, Westview, Boulder, pp.43-86

Short, J. (1981) 'Defence spending in UK regions', Regional Studies, vol. 15, pp.101-110

Simonson, G. R. (1964) 'Missiles and creative destruction in the American aircraft industry, 1956-61', Business History Review, vol. 38, pp.302-14

Smith, D. (1980) The defence of the realm, Croom Helm, London

Smith, M. R. (1985) Military enterprise and technological change: perspectives on the American experience, MIT Press, Cambridge, Mass

Sobel, R. (1972) The age of giant corporations: a microeconomic history of American business 1914-1970, Greenwood, Westport, Conn

Soppelsa, J. (1980) Géographie des armements, Masson, Paris

Southwood, P. (1985) 'The UK defence industry: characteristics of the main UK defence equipment manufacturers which are also relevant to a credible arms conversion strategy', Peace Research Reports No 8, Bradford University School of Peace Studies, September

Stekler, H. O. (1965) The structure and performance of the aerospace industry, University of California Press, Berkeley

Sunseri, A. R. (1977) 'The military-industrial complex in Iowa' in B. F. Cooling (ed.), War, business, and American society: historical perspectives on the military-industrial complex, Kennikat, Port Washington, NY, pp.158-70

Swan, T. G. (1985) 'Nimrod AEW: an airborne control centre' in Advances in command, control and communication systems: theory and applications, Institution of Electrical Engineers, London, pp.220-3

Tapper, O. (1973) Armstrong Whitworth aircraft since 1913, Putnam, London

Tiebout, C. M. (1966) 'The regional impact of defense expenditures: its measurement and problems of adjustment' in R. E. Bolton (ed.), Defense and disarmament: the economics of transition, Prentice-Hall, Englewood Cliffs, pp.125-39

Tirman, J. (ed.) (1984) The militarization of high technology, Ballinger, Cambridge, Mass

Todd, D. (1985) The world shipbuilding industry, Croom Helm, London

_____ and Humble, R. D. (1987) World aerospace: a statistical handbook, Croom Helm, London

_____ and Simpson, J. (1985) 'Aerospace, the state and the regions: a Canadian perspective', Political Geography Quarterly, vol. 4, pp.111-30

_____ (1986) The world aircraft industry, Croom Helm, London

Trebilcock, C. (1969) ' "Spin-off" in British economic history: armaments and industry, 1760-1914', Economic History Review, vol. 22, pp.474-90

_____ (1973) 'British armaments and European industrialization, 1890-1914', Economic History Review, vol. 26, pp.254-72

_____ (1977) The Vickers brothers: armaments and enterprise 1854-1914, Europa Publications, London

_____ (1981) The industrialization of the Continental powers 1780-1914, Longman, London

Treddenick, J. M. (1984) 'Regional impacts of defence spending' in B. MacDonald (ed.), Guns and butter: defence and the Canadian economy, CISS, Toronto, pp.132-58

Tuomi, H. and Väyrynen, R. (1982) Transnational corporations, armaments and development, Gower, Aldershot

Ullmann, J. E. (1984) 'The Pentagon and the firm' in J. Tirman (ed.), The militarization of high technology, Ballinger, Cambridge, Mass., pp.105-22

Utton, M. A. (1982) The political economy of big

business, St Martin's Press, New York
Vanhecke, C. (1986) 'Brazil breaks into arms sales in a big way', _Manchester Guardian Weekly_, 6 April, p.12
Wainwright, H. and Elliott, D. (1982) _The Lucas plan: a new trade unionism in the making?_, Allison & Busby, London
Walters, B. (1986) 'French naval technology', _Navy International_, vol. 91, October, pp.585-96
Weatherbee, D. E. (1986) 'Indonesia: its defense-industrial complex' in J. E. Katz (ed.), _The implications of Third World military industrialization_, Lexington Books, Lexington, pp.165-85
Weaver, D. C. and Anderson, J. R. (1969) 'Some aspects of metropolitan development in the Cape Kennedy sphere of influence', _Tijdschrift voor Econ en Soc Geografie_, vol. 60, pp.187-92
Wettern, D. (1986) 'The dockyard battle', _Navy International_, vol. 91, December, pp.763-5
White, M. (1985) 'General Dynamics boss quits', _Manchester Guardian Weekly_, 2 June, p.9
Wilkinson, E. (1939) _The town that was murdered: the life-story of Jarrow_, Victor Gollancz, London
Wilson, M. (ed.) (1985) _Jane's avionics 1985-86_, Jane's Publishing, London
Zilbert, E. R. (1981) _Albert Speer and the Nazi Ministry of Arms: economic institutions and industrial production in the German war economy_, Associated University Presses, East Brunswick, NJ